Springer Theses

Recognizing Outstanding Ph.D. Research

Aims and Scope

The series "Springer Theses" brings together a selection of the very best Ph.D. theses from around the world and across the physical sciences. Each volume has been selected for its scientific excellence by an internationally renowned institute or department. For greater accessibility to non-specialists, the published versions include an extended introduction, as well as a foreword by the student's supervisor explaining the special relevance of the work for the field. The series as a whole provides a valuable resource both for newcomers to the research fields, and for other scientists seeking detailed background information on special questions. Finally, it serves as an accredited documentation of the valuable contributions made by today's younger generation of scientists.

How to Apply

Theses prepared at internationally leading universities and institutes may be nominated for publication by supervisors with the endorsement of the head of department. Each such department is entitled to nominate one thesis per year. Enquiries and requests for a nomination form should be sent, together with the thesis, to Ms. Nathalie Berg (see link on right).

Eligibility Criteria

To qualify for consideration, theses should fulfil the following criteria:

- They must be written in good English.
- The topic should fall within the confines of Chemistry, Physics, Earth Sciences, Engineering and related interdisciplinary fields such as Materials, Nanoscience, Chemical Engineering, Complex Systems and Biophysics.
- The work reported in the thesis must represent a significant scientific advance.
- A maximum of 30% of the text may be a verbatim reproduction from the author's previous publications.
- They must have been examined and passed during the 12 months prior to nomination.
- The theses should have a clearly defined structure including an introduction accessible to new PhD students and scientists not expert in the relevant field.
- After acceptance, theses should be supplemented by a "Supervisor's Foreword" outlining the wider scientific relevance of the work.

Indexed by zbMATH.

Alexander Hipp

Searching High and Low for the Axion

Doctoral Thesis accepted by
the University of Florida, Gainesville, USA

Springer

Author
Dr. Alexander Hipp
Department of Physics and Astronomy
Northwestern University
Evanston, IL, USA

Supervisor
David B. Tanner
Department of Physics
University of Florida
Gainesville, USA

ISSN 2190-5053 ISSN 2190-5061 (electronic)
Springer Theses
ISBN 978-3-032-02275-2 ISBN 978-3-032-02276-9 (eBook)
https://doi.org/10.1007/978-3-032-02276-9

This Springer imprint is published by the registered company Springer Nature Switzerland AG
The registered company address is: Gewerbestrasse 11, 6330 Cham, Switzerland

Supervisor's Foreword

This book describes research carried out by Alexander Hipp to search for dark-matter axions. Of all the unsolved problems facing contemporary science, identifying the nature of the dark matter in the Universe is one of the most pressing and most significant for understanding the structure and dynamics of a wide variety of cosmological structures from galaxies to the preponderance of supermassive black holes. Axions are now believed by many to be the most promising candidate for dark matter. That axions are widely viewed this way is proven by the number of new axion experiments that are being mounted worldwide. Until recently, the only experiment sensitive to plausible couplings of the axion to two photons has been the Axion Dark Matter eXperiment (ADMX).

ADMX uses a high-Q microwave cavity in a strong magnetic field to capture photons generated by the decay of axions. The cavity resonant frequency must be tuned to match the total energy of the axion. This unknown energy $\mathscr{E}$ is basically the mc^2 of the axion plus a small kinetic energy $K = mv^2/2$ and the frequency is of course $\mathscr{E}/\mathrm{h}$. The cavity resonance is tuned by a small amount and the cavity output recorded for 100 s. A Fourier transform yields the spectrum, which is then searched for extra emitted photons above the background thermal and instrumental noise. Alex Hipp devised a "multi-resolution" analysis of these data from the ADMX detector, improving how the axion detector's signal is searched. The issue is how to optimize the signal-to-noise ratio in the frequency spectrum from the detector given (1) that there is a fixed time (100 s currently) for the scan, (2) that the cavity frequency step is small enough that each spectrum has partial overlap with those from several previous and subsequent scans, (3) that the cavity resonance may be returned to an already-measured frequency several months later, (4) that the value of the small kinetic energy is not a given, (5) that it is the relative axion-detector velocity that matters, and (6) that the detector is in motion through the galaxy, with daily and annual variations in its velocity.

Working with Aaron Quiskamp, a graduate student from Australia, Alex analyzed the Run 1C data from ADMX, which covered the mass range 3.3–4.2 μeV (frequencies of 0.8–1.02 GHz). It started with a "high resolution" analysis. This analysis is motivated by simulations from Sikivie suggesting that there are unvirialized streams of dark matter axions which pass the Earth's location. All axions in the stream have nearly the same velocity and hence the same relative kinetic energy. (The analogy is that the speeds of the streams of water in a waterfall falling off a cliff and measured part way down the waterfall are all about the same.) The frequencies of the photons from decay of these unvirialized particles would fall in a very narrow range of the RF spectrum, perhaps only a few tens of milliHertz in width. This can be compared to ~Ghz frequency set by the few μeV rest mass and the kHz width of the expected Maxwellian energy spectrum for a virialized component. The project included statistical analysis and accounting for Doppler effects of the Earth's orbital and rotational motion. The axion signal should be persistent but the changing Doppler shifts meant that it occurs at different frequencies when observed at different times. The analysis was also extended to a set of spectral resolutions between 20 mHz and 1 Hz, demonstrating the improved signal-to-noise ratio obtained by his method under the hypothesis that the (unknown) axion kinetic energy spread could have a value larger than 10 mHz.

The University of Florida contributes to the design, construction, and testing of tunable cavities for the ADMX experiment. Studies of cavities are described here; cavities resonating in the 1.4–2 and 2–4 GHz ranges were designed, prototypes built, and their resonant frequencies and quality factors measured. Alex's initial prototype validation led to the manufacture of a four-cavity array to be used in the next year to extend ADMX's search to higher axion masses. He also measured several variations on a design that will be used to extend the ADMX frequency range upwards even further, and has contributed to improvements of at least a factor of two in the quality factor of these cavities.

There is a lower limit to the resonant cavity approach, set by the size of the cavity that would fit in even a big solenoid magnet. For ADMX, this frequency is about 500 MHz, corresponding to an axion mass of 2 μeV. There are theoretical considerations which allow the mass to be below this limit, perhaps by an order of magnitude, or two, or even more. Cavities at these frequencies would be huge, as would be the cost of a magnet to contain them. The approach Alex contributed to is the use of a lumped *LC* circuit. At Florida we have been interested in lower axion masses, and, hence, lower frequency resonators. A former student, Nicole Crisosto, built a prototype [1] called ADMX SLIC (Superconducting LC circuit Investigating Cold axions) and found good performance at low temperatures, with a resonance around 40 MHz. The resonator consists of a single turn of NbTi wire and a parallel plate capacitor. The loop is in the bore of an 8 T magnet. The frequency (or mass) range reported in this paper was limited by a failure of the piezoelectric motor that tuned the capacitor. Alex redesigned the tuning mechanism and other aspects of the experiment, and, upon graduation, passed the apparatus on to another student. A coil couples the flux in the resonator to a cooled low-noise transistor amplifier, the front end of the SLIC receiver. We hope to scan a range of 25–50 MHz (axion masses of 0.1–0.2 μeV). The sensitivity will be above the QCD axion limits, but there is always discovery potential for very light axion-like particles. More importantly this experiment can point the way to a full scale experiment using an ADMX-sized or larger magnet.

Gainesville, FL, USA
June 2025

David B. Tanner

Reference

1. N. Crisosto, J.R. Gleason, P. Sikivie, N.S. Sullivan, D.B. Tanner, J. Yang, and G. Rybka, Phys. Rev. Lett. **124**, 241101 (2020).

Acknowledgements I would like to thank my advisor Dr. David B. Tanner for his excellent guidance, which made my doctoral journey a smooth and enjoyable experience. I want to thank Dr. Shriram Jois for introducing me to the UF axion group and his encouragement to pursue it for my Ph.D. My first time meeting the entire UF axion group was a celebratory goodbye dinner for Dr. Jihee Yang. Despite being a stranger to several members, I was welcomed very warmly by all, for which I am incredibly appreciative. Thank you to Dr. Pierre Sikivie and Dr. Neil Sullivan for your assistance and mentoring. I thank Joe Gleason for significant assistance in all engineering aspects of my project. I would also like to thank Dr. Y. K. Yoon for his support on my supervisory committee and helpful comments. For their guidance and wonderful discussions on analysis, I thank Dr. William Wester, Dr. Leanne Duffy, Dr. Michael Tobar, and Aaron Quiskamp. I would like to thank Dr. Gp Carosi, Dr. Gray Rybka, and every other member of the ADMX collaboration that make this axion search possible. I could not have picked a better or more welcoming team to join. I would like to thank the Institute for High Energy Physics and Astrophysics for awarding me the IHEPA Graduate Fellowship, funding my research during the summer of 2020, as well as the DOE under grant No. DE-SC0010296, through which I was supported for all four years of my research. Last but not least, I would like to thank my beautiful wife Emily for her support throughout the entire process.

Competing Interests The author has no competing interests to declare that are relevant to the content of this manuscript.

Parts of this thesis have been published in the following journal articles

A. T. Hipp et al., "Search for non-virialized axions with 3.3–4.2 μeV mass at selected resolving powers", arXiv.2410.09203 (2024) (In review).

Contents

Abbreviations

ADMX	Axion Dark Matter eXperiment
CMB	Cosmic Microwave Background
CP	Charge-Parity
DAQ	Data Acquisition
DM	Dark Matter
EDM	Electric Dipole Moment
FFT	Fast Fourier Transform
FWHM	Full Width at Half Maximum
GUT	Grand Unified Theory
IF	Intermediate Frequency
JPA	Josephson Parametric Amplifier
LO	Local Oscillator
QCD	Quantum Chromodynamics
SA	Spectrum Analyzer
SM	Standard Model of particle physics
SNR	Signal-to-Noise Ratio
VNA	Vector Network Analyzer

Chapter 1
Introduction

Abstract This briefly outlines the work to be presented in this thesis.

Two long-standing problems in physics are the nature of dark matter and the strong CP problem. A possible solution to both dark matter and the strong CP problem is a hypothetical particle called the axion [1–3].

The primary interaction that experiments exploit to detect the axion is its coupling to photons [4–6].

$$\mathscr{L}_{a\gamma\gamma} = -\frac{1}{4} g_{a\gamma\gamma} a F_{\mu\nu} \tilde{F}^{\mu\nu}, \tag{1.1}$$

where a is the axion field, and $F_{\mu\nu}$ and $\tilde{F}^{\mu\nu}$ are the electromagnetic tensor and its Hodge dual, respectively, and $g_{a\gamma\gamma}$ is a coupling constant. An interaction described by Eq. 1.1 is the decay of an axion to two photons, or when the axion passes through a strong magnetic field, to a single photon. The energy of the photon is given by $hf = \frac{1}{2} m_a^2 v^2 + m_a^2 c^2$, where m_a is mass of the axion, v the axion velocity, and f the photon frequency. These photons can be resonantly enhanced by a microwave cavity if f is approximately equal to the frequency of a cavity mode. Experiments such as ADMX, HAYSTAC, and DM RADIO utilize tunable resonators to scan a range of axion masses [7–9]. Two projects have been completed, which have added to the axion search effort. A multi-resolution analysis, sensitive to alternative dark matter halo models and frequency modulations, was carried out on the ADMX run 1c dataset. Second, the early stages of prototyping a future ADMX Experiment, ADMX Extended Frequency Range, was completed. The results of each are presented in this work.

A. Hipp, *Searching High and Low for the Axion*,
Springer Theses, https://doi.org/10.1007/978-3-032-02276-9_1

References

1. R.D. Peccei, H.R. Quinn, Phys. Rev. Lett. **38**, 1440 (1977). https://doi.org/10.1103/PhysRevLett.38.1440
2. S. Weinberg, Phys. Rev. Lett. **40**, 223 (1978). https://doi.org/10.1103/PhysRevLett.40.223
3. F. Wilczek, Phys. Rev. Lett. **40**, 279 (1978). https://doi.org/10.1103/PhysRevLett.40.279
4. P. Sikivie, Phys. Rev. Lett. **51**, 1415 (1983). https://doi.org/10.1103/PhysRevLett.51.1415
5. P. Sikivie, Phys. Rev. D **32**, 2988 (1985). https://doi.org/10.1103/PhysRevD.32.2988
6. J. Ipser, P. Sikivie, Phys. Rev. Lett. **50**, 925 (1983). https://doi.org/10.1103/PhysRevLett.50.925
7. C. Bartram et al., Phys. Rev. Lett. **127**, 261803 (2021). https://doi.org/10.1103/PhysRevLett.127.261803
8. L. Zhong et al., Phys. Rev. D **97**, 092001 (2018). https://doi.org/10.1103/PhysRevD.97.092001
9. L. Brouwer, DMRadio Collaboration, Phys. Rev. D **106**, 112003 (2022). https://doi.org/10.1103/PhysRevD.106.112003

Chapter 2
Background

Abstract Two of the biggest mysteries in physics are the strong CP problem in particle physics and the unknown nature of dark matter. Despite its successes, the SM cannot explain either of these and thus new physics is required. In particular, a new pseudoscalar boson called the axion has the potential to resolve both mysteries. This chapter introduces and motivates the significance of the strong CP problem and dark matter, describes how the axion can solve each, and discusses how the axion can be detected. The detection portion primarily focuses on the Axion Dark Matter eXperiment (ADMX). The axion is a well-motivated solution to the dark matter and strong CP problem. The evidence for the axions existence as a fundamental particle and as dark matter, and methods for detection are discussed.

2.1 Axions

2.1.1 The $U(1)_A$ Problem

During the 1970s, a problem regarding $U(1)_A$, the axial-vector symmetry of QCD, was discovered [1, 2]. All formulations of QCD have an axial-vector current, J_5^μ, associated with the $U(1)_A$ symmetry, which is not conserved. Initially, this axial-vector symmetry was assumed to be a true symmetry of QCD, which is spontaneously broken. Goldstone's theorem states the spontaneous breaking of a continuous, global symmetry implies the existence of a Goldstone boson [3]. The broken $U(1)_A$ symmetry suggested the existence of an additional pseudoscalar boson in the SM. This particle was quickly ruled out and Weinberg dubbed this mystery the $U(1)_A$ problem [4].

A solution to the $U(1)_A$ problem was first proposed by 't Hooft [5, 6]. He showed the divergence of the axial-vector current acquires quantum corrections due to the Adler-Bell-Jackiw anomaly, computed using the triangle graph [7, 8]. The divergence of the axial-vector current with quantum corrections is given by

$$\partial_\mu J_5^\mu = \frac{g_s^2 N}{32\pi^2} G_a^{\mu\nu} \tilde{G}_{a\mu\nu}, \tag{2.1}$$

A. Hipp, *Searching High and Low for the Axion*,
Springer Theses, https://doi.org/10.1007/978-3-032-02276-9_2

where g_s is the strong coupling constant, $G_a^{\mu\nu}$ is the gluon field strength tensor, and $\tilde{G}_{a\mu\nu} = \frac{1}{2}\varepsilon_{\mu\nu\alpha\beta}G_a^{\alpha\beta}$. The chiral anomaly results in a non-zero contribution to the action, only with a proper choice of gauge configuration, as shown by Hooft [5, 6]. Choosing a correct gauge shows $U(1)_A$ is not a symmetry of QCD, however the consequence is the QCD vacuum structure is more complex than initially thought.

The true vacuum is the θ-vacuum, a superposition of n-vacua [2]. Computing the amplitude of the transition from one vacuum configuration to another, one finds the QCD Lagrangian acquires an additional term, given by [2]

$$L_\theta = \theta \frac{g_s^2}{32\pi^2} G_a^{\mu\nu}\tilde{G}_{a\mu\nu}, \tag{2.2}$$

where $0 \leq \theta \leq 2\pi$.

2.1.2 The Strong CP Problem

The additional term in the QCD Lagrangian, given by Eq. 2.2, violates CP symmetry because the Levi-Civita tensor is CP odd, while the gluon field strength tensor is CP even. This in and of itself is not an issue. CP symmetry is known to be an approximate symmetry and its violation has been observed in weak interactions. An experiment studying the beta decay of cobalt-60 by Wu and collaborators was the first to show weak interactions violate parity symmetry [9]. A later experiment provided evidence for indirect CP violation in the decay of kaons [10]. Since then, many experiments have observed direct CP violation, such as in the decays of B mesons and D mesons [11, 12]. Given these observations, one may expect to see CP violations pop up in strong interactions as well.

In particular, the CP violating term results in the neutron acquiring a rather large electric dipole moment, proportional to θ, given by [13, 14]

$$|d_N| \approx \theta \times 10^{-16}\text{e cm}, \tag{2.3}$$

where d_N is the neutron EDM. A recent upper-bound on the neutron EDM is 3×10^{-26} e cm, placing an upper-bound $\theta \lesssim 10^{-10}$ [15]. This finding suggests θ may actually be zero. The smallness of θ is known as the strong CP problem.

When weak interactions are included in the Lagrangian,the quark mass matrix, M, is generally complex. To transform to a basis where M, and thus the quark masses are real, a chiral transformation must be done. This shifts θ to $\bar{\theta} = \theta + argdet M$, where $\bar{\theta}$ is the physical observable [2].

Several solutions to the strong CP problem have been proposed. An incredibly unsatisfactory one is that $\bar{\theta}$ just happens to be zero. Another requires the up quark to have zero mass, however early measurements of the up quark mass suggests this to be unlikely [16]. In 1977, Peccei and Quinn proposed a dynamic solution to the strong CP problem[17].

2.1.3 Peccei-Quinn Solution

Peccei and Quinn introduced a new global, chiral $U(1)_{PQ}$, with a chiral anomaly like $U(1)_A$, which is spontaneously broken. The SM Lagrangian is modified by the addition of a new PQ-field, given by [2]

$$\mathcal{L} = \mathcal{L}_{QCD} + \theta \frac{g_s^2}{32\pi^2} G_a^{\mu\nu} \tilde{G}_{a\mu\nu} - \frac{1}{2}\partial_\mu a \partial^\mu a + \xi \frac{a}{f_a} \frac{g_s^2}{32\pi^2} G_a^{\mu\nu} \tilde{G}_{a\mu\nu} + \mathcal{L}_{int}(a/f_a, \psi), \tag{2.4}$$

where a is the PQ field, ξ is a constant, and $\mathcal{L}_{int}(a/f_a, \psi)$ are axion-fermion interactions. Peccei and Quinn showed the minimum energy solution occurs at $\langle a \rangle = -\bar{\theta}\frac{f_a}{\xi}$, so the CP violating term gets canceled out [17]. This solution provides a dynamic argument for the non-observation of CP violation in strong interactions.

After Peccei and Quinn published their results in 1977, six months later Weinberg and Wilczek released papers showing that oscillations of the PQ-field about its minimum would produce a new pseudoscalar boson, (as expected by Goldstone's theorem), named the axion [1, 18]. In Peccei, Quinn, Weinberg, and Wilczek's original model, the symmetry breaking scale was assumed to be on the order of the electroweak scale, resulting in an axion with a mass of approximately keV to Mev. However, this was quickly ruled out by beam dump experiments and investigation of various nuclear decays in which the axion would participate [19, 20].

2.1.4 Benchmark Models

It was quickly realized an axion with a small mass and very weak coupling is not ruled out the initial searches. The first of these was proposed by Kim and Shifman, Vainshtein, and Zakharov, now called the KSVZ model, in which the axion and a new heavy quark are added to the SM [21, 22]. In this case, the axion only couples to quarks in the SM. The second model, called the DFSZ model after Dine, Fischler, Srednicki, and Zhitnitsky, adds an axion and two Higgs doublets to the SM model [23, 24]. DFSZ axions would couple to both quarks and leptons. The KSVZ and DFSZ axion are often called the QCD axion. Although the KSVZ and DFSZ models do not make predictions about the axion mass, they do provide relationships between its mass and coupling to nucleons and photons, the latter of which is discussed in Appendix A.

All axion searches today set limits on the axion coupling to either nucleons or photons, versus its mass. The predictions made by KSVZ and DFSZ have become the benchmark to reach in experimental sensitivity.

2.1.5 The Axion Mass

In the KSVZ and DFSZ models, the axion mass can be written as [1, 2]

$$m_a = \frac{\nu_F}{f_a}\bar{m}_a, \tag{2.5}$$

where $\nu_F \simeq 250\,\text{GeV}$ is the electroweak scale, f_a is the PQ symmetry breaking scale, and $\bar{m}_a$ is the axion mass due to mixing with the pion given by

$$\bar{m}_a = m_\pi \frac{f_\pi}{\nu_F}\frac{\sqrt{m_u m_d}}{m_u + m_d}, \tag{2.6}$$

where m_π is the pion mass, f_π is the pion decay constant, and m_u and m_d are the up and down quark masses, respectively. The values of these parameters are

$$m_u \approx 2.3 \times 10^{-3}\ \text{GeV} \tag{2.7}$$

$$m_d \approx 4.8 \times 10^{-3}\ \text{GeV} \tag{2.8}$$

$$m_\pi \approx 0.140\ \text{GeV} \tag{2.9}$$

$$f_\pi \approx 0.093\ \text{GeV} \tag{2.10}$$

Substituting these into Eq. 2.6, then this value into Eq. 2.5 yields

$$m_a \simeq 6\mu\text{eV}\left(\frac{10^{12}\text{GeV}}{f_a}\right). \tag{2.11}$$

The symmetry breaking scale is unknown, however various models make predictions about f_a, such as axions embedded in GUTs and string theory [25, 26]. In addition, some cosmological models make predictions about what mass range the axion can live in, which will be discussed in a later section.

2.1.6 Astrophysical Constraints

The axion parameter space has been constrained through a combination of experimental searches and astrophysical observations. Figure 2.8 shows constraints on the axion's interaction with two photons. The observation of SN 1987A was used to placed stringent limits on the axion [27]. Supernovae result in the release of an immense number of neutrinos and axions. Neutrinos from SN 1987A were detected by the Kamiokande II and Baskan neutrino observatories. If axions were too heavy or had a coupling to nucleons within a certain range, they would have been observed by the water Cherenkov detected, changing the measured neutrino flux. The axion-nucleon coupling was constrained to $10^{-10} \lesssim g_{aN} \lesssim 1.4 \times 10^{-6}$. Axions with a cou-

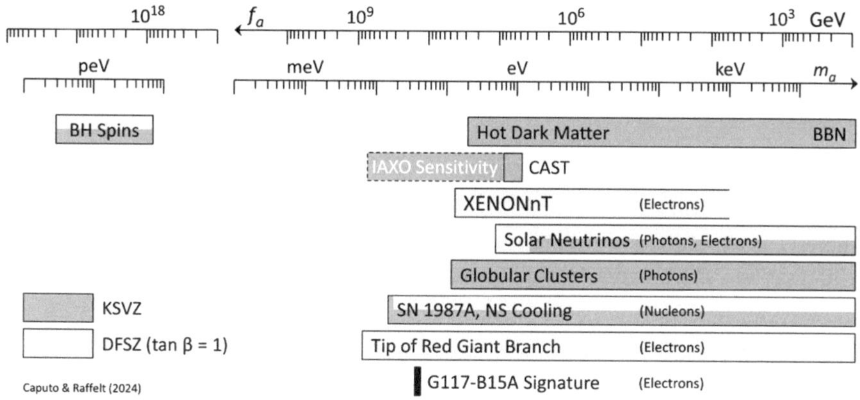

Fig. 2.1 Summary of constraints placed on the axion mass by various astrophysical phenomena. The interaction channel is specified for each. Caputo and G. Raffelt, "Astrophysical Axion Bounds: The 2024 Edition," (2024), arXiv:2401.13728

pling below the lower bound would have free streamed through the detector, resulting in no signal, while axions with a coupling above the upper bound would have remained trapped in the star.

Iwamota showed nucleon bremsstrahlung, $NN \rightarrow NNa$, could contribute significantly to the cooling of neutron stars [28]. Additional analysis of SN 1987A and observations of a few neutron stars constrained the axion mass to $m_a < 15$ meV for KSVZ axions and $m_a < 19$ meV for DFSZ axions [29]. Analysis of various other astrophysical observations, such as the evolution of globular-cluster stars and white dwarf cooling place similar bounds [29]. Figure 2.1 shows a summary of astrophysical constraints on the axion mass. These observations exclude heavy axions.

2.2 Dark Matter

The dark matter problem arises from discrepancies between measurements made in the last century and theoretical predictions. The theory of dark matter is quite compelling as there is evidence for it at the smallest and largest scales of the universe. Solutions to the dark matter problem do not replace the currently accepted theories (e.g. general relativity and the standard model), but rather are modifications.

2.2.1 *Galaxy Dynamics*

One of the first theories to make precise predictions about the dynamics of large scale structures, such as the solar system, was Newton's law of gravity. A key prediction of this theory is relationship between the orbital radius of a body and its orbital velocity, about a much heavier object. Setting the acceleration due to gravity equal to the centripetal acceleration of a rotating body yields

$$\frac{GM}{r^2} = \frac{v^2}{r}, \tag{2.12}$$

where G is the gravitational constant, M is the mass of the central object, r is the orbital radius, and v is the orbital velocity. Solving for velocity gives

$$v = \sqrt{\frac{GM}{r}}, \tag{2.13}$$

so, the orbital velocity of an object is inversely proportional to the square root of its orbital radius. This prediction is correct for the planets in solar system, as shown in Fig. 2.2. Increasing the scale from the solar system level to the galactic level, one would expect this relationship to be approximately correct. The galactic situation is slightly different, as the mass of the solar system is concentrated almost entirely in the sun, where as galaxies have a central bulge where the mass increases significantly up to some critical radius, then falls off [30]. This yields a modified rotation curve, where the orbital velocity of objects increases with radius until a critical point, then falls off according to Eq. 2.13, shown in Fig. 2.3 [30].

Instead what is observed is an increasing orbital velocity to the critical radius, then a region where the velocity plateaus. The divergence from the relationship described by Eq. 2.13 was first observed by Zwicky in the 1930s. He performed a study on the Coma galaxy cluster, measuring the masses of galaxies using their luminosities and

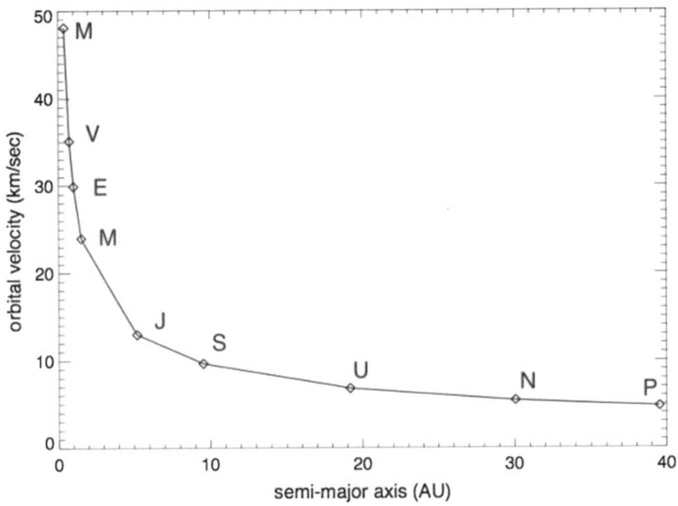

Fig. 2.2 Orbital velocity versus radius for stars in the solar system. Credit: Chris Palma

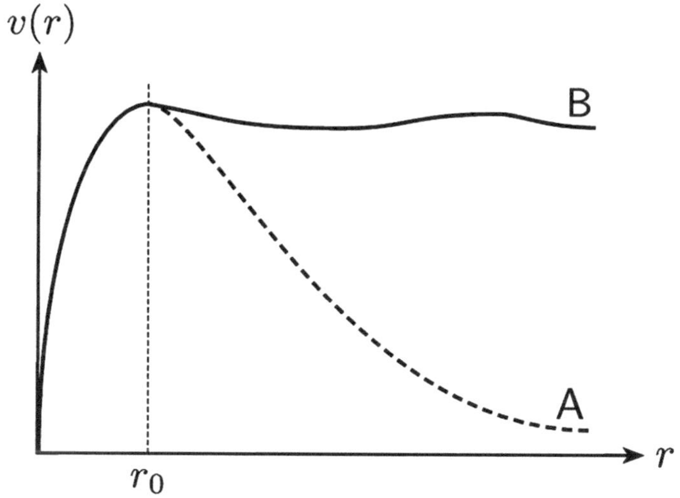

Fig. 2.3 A: Expected galactic rotation curve versus; B: Approximation of observed galactic rotation curves. A. Araujo, D. F. López, and J. G. Pereira, Gravit. Cosmol. 25, 157 (2019)

the virial theorem [31]. For the solar system, 99.9% of the mass is from luminous matter (i.e. the sun) [32]. Assuming stars are the primary contribution to the matter distribution of galaxies and that the previous fact holds true for every star system, it follows that computing the mass of a galaxy using its luminosity should result in an accurate value. However, that is not what Zwicky found. The virial theorem method calculates the mass using the motion of the stars within the galaxies. The orbital velocity of an object is dependent on the total mass within its orbit, independent of the properties of the matter. With the virial theorem method, Zwicky calculated a mass approximately 500 times larger than that given by the luminosity estimate. This suggested the majority of the mass in a galaxy is nonluminous, thus the name dark matter.

This study was followed up during the 1970s by Rubin and collaborators with their survey of 21 galaxies [33]. For each galaxy, they measured the orbital velocity of stars and plotted it versus their radius. The study found orbital velocities increase quickly to some critical radius, as predicted, but then plateau, shown in Fig. 2.4. Their findings suggests the presence of a significant amount of matter beyond the luminous central bulge. This study confirmed Zwicky's results and added to the growing evidence of dark matter.

2.2.2 *Galaxy Collisions*

The evidence for dark matter given by Zwicky and Rubin relies on the current model of gravitational interactions being correct, which some theorists argue is not the case. Observations of gravitational lensing provides more evidence for dark matter,

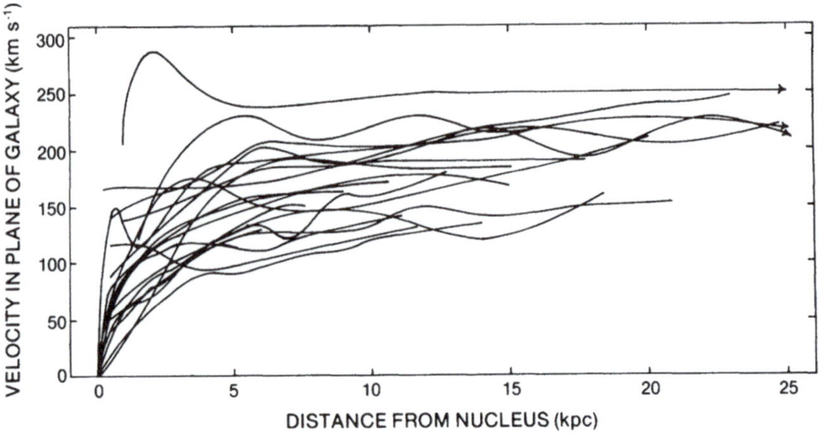

Fig. 2.4 Galactic rotation curves. V. C. Rubin, N. Thonnard, and W. K. Ford, Jr., ApJ 238, 471 (1980)

which can not be explained by modified orbital dynamics. A study by Clowe and collaborators analyzed observations from the collision of two galaxy clusters, often referred to as the Bullet Cluster [34]. They calculated the total mass distribution using observations of weak gravitational lensing and the mass distribution of the baryonic matter using x-ray emission observations. The results showed an 8-sigma spatial offset between the two mass distributions, as shown in Fig. 2.5.

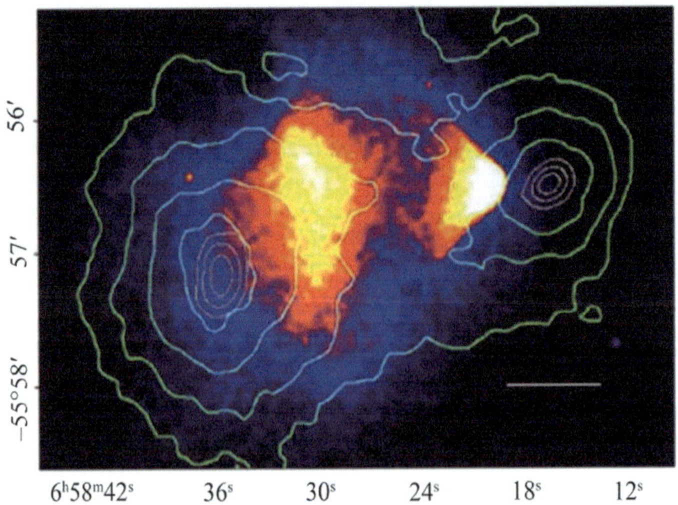

Fig. 2.5 Image showing a contour plot of the total mass distribution from gravitational lensing overlaid a color gradient of the baryonic matter distribution. D. Clowe et al., ApJ 648, L109 (2006)

This result supports a model where galaxy clusters are composed of collisionful hot gas composed of baryonic matter, approximately collisionless stellar components, and collisionless dark matter [35]. The majority of the baryonic mass is in the hot gas. When the galaxy clusters collide, the dark matter and stellar contributions are slowed a little due to gravitational interactions, but continue approximately on their original trajectory. The hot gas instead experiences significant friction and lags behind the previous two components. This study provides significant evidence for the existence of particle dark matter.

2.2.3 CMB Anisotropy

More evidence for dark matter comes from measurements of the CMB [36]. A plot of the CMB temperature power spectrum is shown in Fig. 2.6, where the peak heights and locations are dependent on a number of parameters. Three in particular are the matter density, Ω_m, the baryonic matter density, Ω_b. and the dark energy density, Ω_Λ, which has the relationship $\Omega_\Lambda + \Omega_m = 1$. Changes in these parameters affect the power spectrum in different ways. For example, varying the baryon density will change the relatives heights of the even and odd peaks [37].

The Planck collaboration performed a careful analysis of the CMB temperature power spectrum [38] and found values of $\Omega_m = 0.321$ and $\Omega_\Lambda = 0.679$. The matter density has been measured using other methods, such as cluster evolution and the mass-to-light ratio, which find a similar value [39]. The physical baryon density

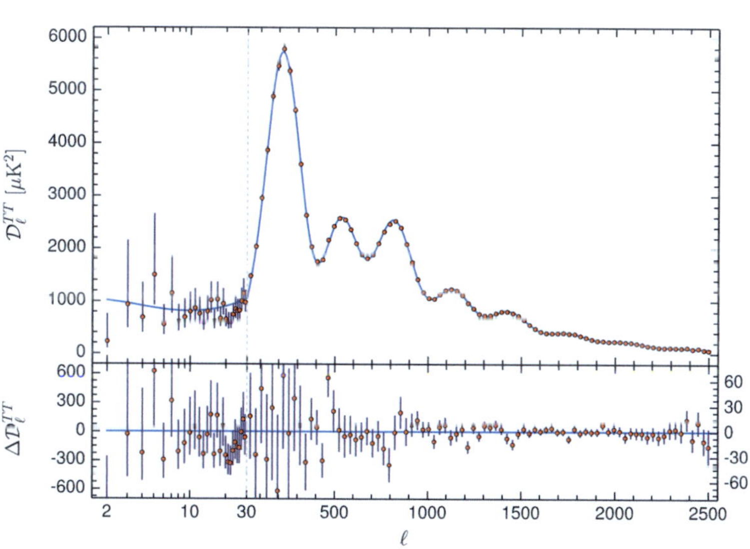

Fig. 2.6 CMB temperature power spectrum. N. Aghanim, Y. Akrami, and all, A & A Volume 641 (September 2020), https://doi.org/10.1051/0004-6361/201833910e

and matter density are given by $\Omega_m h^2$ and $\Omega_b h^2$, where h is the dimensionless Hubble constant. These were calculated to be $\Omega_m h^2 = 0.1434$ and $h^2\Omega_b = 0.02212$, meaning baryonic matter accounts for only 15% of the total matter density and the remaining 85% is dark matter. The baryon matter density has been computed independently using the deuterium to hydrogen number ratio [36]. This ratio is found through observations of deuterium and hydrogen absorption lines. This method finds a baryon matter density of $h^2\Omega_b = 0.0214$, in agreement with the above value.

Unlike the large scale measurements, the anisotropies in the CMB temperature power spectrum are due to quantum fluctuations. This reveals two things about the nature of dark matter: first, dark matter is composed of particles, and second, they were produced in abundance prior to the time of last scattering.

2.2.4 *Dark Matter Candidates*

To be a candidate for dark matter, a particle needs to satisfy four key conditions. First, the non-gravitational interactions between dark matter particles and SM particles must be very feeble. Second, the particle must be produced abundantly in the early universe. This is required to satisfy the significant fraction of the matter density dark matter accounts for at the time of last scattering. Third, the dark matter must be produced cold, that is the particles are non-relativistic. This requirement ensures the particles can form the gravitationally bound states observed today. Lastly, the dark matter particles must be long-lived.

Over the past several decades, many models have been proposed as solutions to the dark matter problem. A few noteworthy candidates are sterile neutrinos, axions, and WIMPS [40]. The first sterile neutrino model was proposed by Dodelson and Widrow, in which an additional neutrino, which does not participate in electroweak interactions, is added to the SM [41]. This feebly acting particle would only couple to the SM through mixing with the three known neutrino species. Although Dodelson and Widrow's original model is ruled out, other sterile neutrino models remain as viable candidates [40].

For any dark matter candidate, its relic abundance must match the dark matter density. The thermal relic abundance is dependent on the self-annihilation cross-section. Taking the known dark matter density and solving for a cross-section yields a value that is incredibly close to that which arises from the weak force [37, 40]. This finding propelled models of Weakly Interacting Massive Particles (WIMPS) to the forefront of the search for dark matter. From several decades ago to today, WIMPS have remained the leading and most intensely searched for dark matter candidate.

Particle collider experiments, such as ATLAS and CMS at CERN, have been able to place some limits on the WIMP parameter space, typically by searching for missing momentum [42]. However, the primary search method utilizes time projection chambers, filled with a liquid such as xenon, and aims to detect the recoil energy from a collision between WIMP particles and the atoms in the liquid [43–46]. These experiments place limits on the WIMP-nucleon interaction cross-section versus the WIMP mass, where the mass is in the range 10 GeV to 1,000 GeV.

The most recent results by the LUX-ZEPLIN (LZ) collaboration place most stringent limits to date on the WIMP parameter space [47]. WIMP experiments continue to improve their sensitivity; however, they are fast approaching difficult territory. At a certain sensitivity, the experiments will begin to detect solar neutrinos, which will create a significant background [48]. Often referred to as the neutrino fog, once this region of parameter space is reached, the SNR of a WIMP signal becomes incredibly difficult to improve. With no WIMP detection over a few decades of data taking and a challenging road ahead, motivation for an alternative dark matter candidate has been increasing.

The second most popular dark matter candidate today is the axion, however its support is growing fast as WIMP skepticism increases.

2.2.5 Alternative Theories

Although particle dark matter is the most supported model by the physics community, some theorists propose alternatives. One is Modified Newtonian Dynamics (MOND) [49–51]. The essential idea of these theories is that in the limit of small acceleration, a modified version of Newton's law of gravitation is required. They argue the seemingly missing mass in galaxies is due to our understanding of gravity being incorrect on such large scales. However, these theories often fail to explain evidence naturally explained by particle dark matter. For example, observations of gravitational waves provide evidence against MOND as a dark matter alternative [52, 53]. Although MOND may provide explanations for other phenomena, it is unlikely to be the solution to the dark matter problem.

A brief mention of one other alternative, a theory of Emergent Gravity [54]. This model argues that if gravity is an emergent phenomenon, then there will exist an additional dark gravity force, which can explain the evidence for dark matter. It is important to note that Emergent Gravity is distinctly different from MOND. Unlike MOND, which is an extension of the current theory of gravity, Emergent Gravity begins with first principles and shows the gravitational laws that emerge contain additional properties.

2.2.6 Axion Dark Matter

Recall the requirements a particle must meet in order to be a viable dark matter candidate: interacts feebly with SM particles, long-lived, produced in abundance during the early universe, and produced non-relativistically.

The QCD axion discussed earlier describes an axion which satisfies the first two conditions. The last two properties are dependent on the cosmology of the axion.

The axion potential is 2π periodic and when symmetry breaking of $U(1)_{PQ}$ occurs, it is accompanied by a broken $Z[N]$ symmetry, where N is the number of quark flavors which rotate under $U(1)_{PQ}$ [55, 56]. The minimum value the axion field takes on is dependent on whether the symmetry breaking occurs before or after inflation.

Prior to inflation, the universe was contained in a single causally connected patch, so if symmetry breaking occurs before inflation, the axion field will take on a single minimum value throughout all of space.

In the event the symmetry breaking occurs after inflation, causally disconnected patches of the universe may have different values for the axion ground state, resulting in domain walls [55]. This scenario introduces additional production mechanisms for the axion.

2.2.7 Axion Production in the Early Universe

How axions are produced during the early universe depends on the time of symmetry breaking with respect to inflation [2]. First, consider the case of symmetry breaking occurring prior to inflation.

The only process that would produce cold axions is vacuum realignment. The axion field starts with some amount of potential energy, then as it begins to oscillate about its minimum, energy is transferred to the production of particles. Only the zero-momentum mode contributes to this process.

Now, consider the case of symmetry breaking post inflation. Vacuum realignment contributes in this case as well, however there are also contributions from non-zero momentum modes [2]. Two additional sources are the decay of axion strings and domain walls [55, 57].

In either case, axions are capable of account for the entirety of the dark matter density, and thus being the primary constituent of galactic dark matter halos, if $f_a \approx 10^{12}$ GeV [58–60].

2.3 Axion Detection

Most axion experiments aim to detect axions from the dark matter halo in the Milky Way galaxy. The primary interaction exploited by these experiments is the axion's coupling to two photons. Although the axion is capable of decaying to two photons in vacuum, this process is incredibly rare, with a lifetime 10^{27} years $\lesssim \tau_a \lesssim 10^{42}$ years, which is significantly longer than the age of the universe [2]. However, the decay can be stimulated by introducing an external magnetic field. When an axion passes through a magnetic field, it can couple to a virtual photon and decay to a single photon. These processes are shown in Fig. 2.7.

The coupling between axions and photons is often referred to axion electrodynamics, as the equations of motion are simply Maxwell's equations modified by an

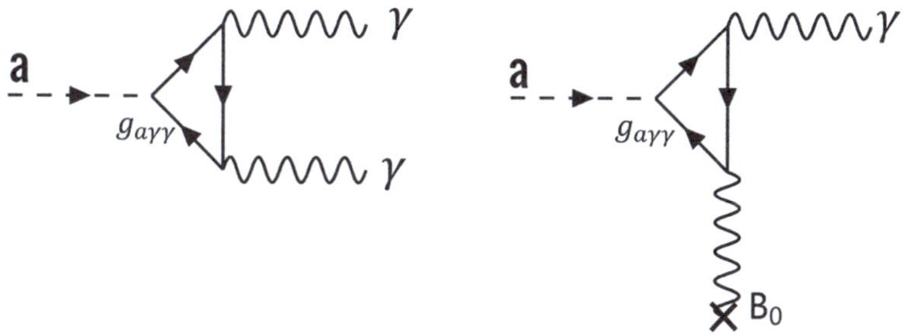

Fig. 2.7 Decay of an axion to a photon; Left: In vacuum; Right: Axion passing through an external magnetic field

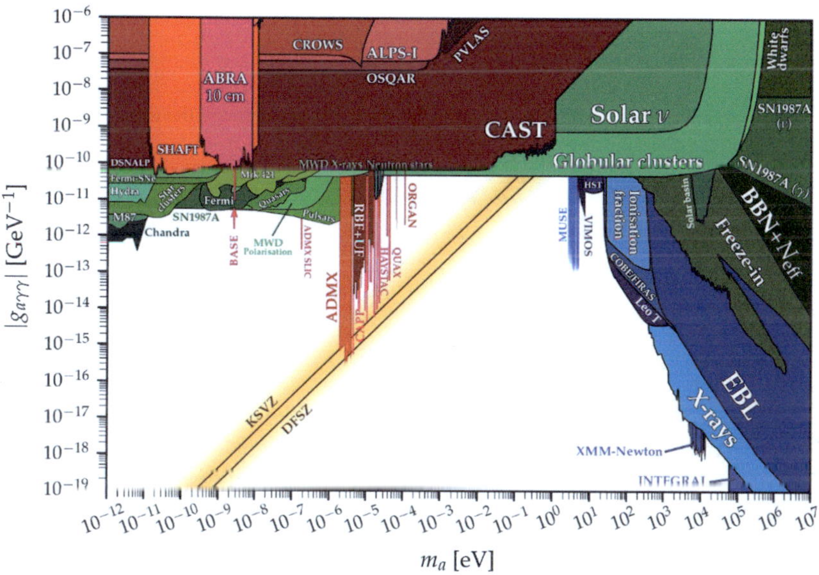

Fig. 2.8 Exclusion plot for the axion-photon-photon coupling versus the axion mass. The limits are from a combination of experimental and astrophysical observations

axion contribution. The equations describing the axion-photon-photon interaction are discussed in Appendix A. Current limits on the axion-photon-photon coupling are shown in Fig. 2.8. Sikivie was the first to design an experiment based on this process, called the Sikivie haloscope [61, 62].

2.3.1 *Axion Haloscopes*

The Sikivie haloscope consists of a microwave cavity inside a magnetic field [61, 62]. When an axion passes through the magnetic field and inside the cavity, it has a small

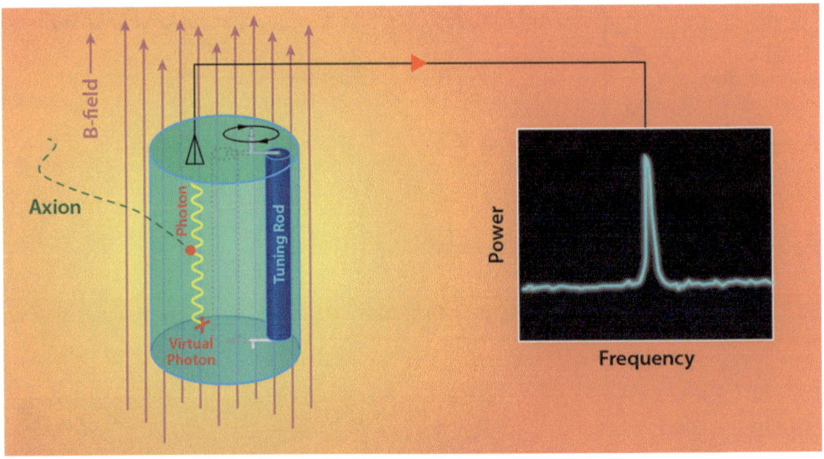

Fig. 2.9 Diagram of an axion haloscope, with an axion passing through the cavity and the signal being read out to produce its power spectrum

probability to decay to a microwave photon with energy

$$E_\gamma = hf = m_a c^2 + \frac{1}{2} m_a v_a^2, \tag{2.14}$$

where h is Planck's constant, f is the photon frequency, and v_a is the axion velocity, with respect to the detector, which is determined by the dark matter halo dynamics. When the frequency of the produced photon is approximately equal to the frequency of a cavity mode, a resonance can build up [62]. A diagram of this process is shown in Fig. 2.9.

Many axion experiments, such as ADMX, utilize a solenoid magnet. In cylindrical coordinates, with $\vec{B}_0 = B_0\hat{z}$, Equations A.8 and A.9 show the effective axion current will be in the $\hat{z}$-direction, resulting in an axion magnetic field in the $\hat{\phi}$-direction.

A carefully designed resonator can be driven by the transfer of energy from the axion field to EM waves when placed inside the solenoid magnet. The most common resonator utilized is a right circular cylindrical resonant cavity. The axion field can pump energy into cavity TM modes with a frequency equal to that given by Eq. A.10.

The power of an axion signal in a cavity haloscope is dependent on a number of parameters. Ultimately, what an experiment measures is the power of the produced photons. Figure 2.7 shows the decay of one axion produces a single photon, so the power will simply be the product of the number of axions in the cavity, the probability of decay, the energy of the resulting photon, and the ability of the cavity to support a resonance. The number of axions in the cavity is given by

$$N_a = (V C_{lmn}) \frac{\rho_a}{m_a}, \tag{2.15}$$

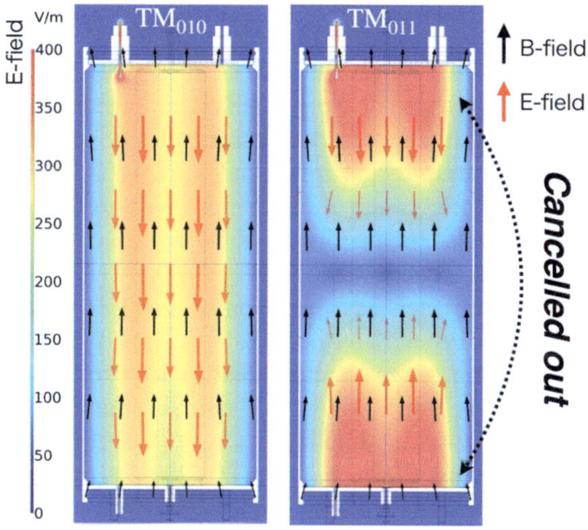

Fig. 2.10 A comparison of the electric fields for the TM_{010} and TM_{011} cavity modes

where V is the cavity volume, ρ_a is the local axion mass density, and C_{lmn} is a mode dependent form factor given by

$$C_{lmn} \equiv \frac{|\int d^3x \vec{B}_0 \cdot \vec{E}_{lmn}(\vec{x})|^2}{V * B_0^2 \int d^3x \varepsilon(\vec{x}) |\vec{E}_{lmn}(\vec{x})|^2}, \tag{2.16}$$

where $\vec{E}_{lmn}(\vec{x})$ is the electric field of the lmn cavity mode and $\varepsilon(\vec{x})$ is the permittivity within the cavity. The form factor effectively describes the overlap of a cavity mode and the axion field, so the product of the form factor and volume yields an effective volume. Figure 2.10 depicts the electric field for the TM_{010} and TM_{011} modes.

The TM_{010} mode has a large form factor, since its electric field points in the same direction along the entire length of the cavity. The electric field of the TM_{011} mode is symmetric about the midpoint of the cavity, so integrating its dot product with the B-field over the volume yields a form factor of zero. The TM_{010} mode has largest form factor of any cavity TM mode, so it is the one used for axion detection.

The decay probability can be written as

$$\sigma \propto (g_{a\gamma\gamma}^2 B^2) \frac{1}{m_a}, \tag{2.17}$$

where $g_{a\gamma\gamma}$ is a model dependent coupling, B is the strength of the magnetic field, and m_a is the axion mass. The ability of a cavity to support a resonant mode is given by its loaded quality factor, Q_L, which is described in Appendix B. However, if the quality factor of the axion flow, Q_a, is smaller than Q_L, then the power will be limited by Q_a. Putting these together yields

$$P_{a,lmn} = (g_{a\gamma\gamma}^2 B^2)\frac{1}{m_a}(V \times C_{lmn})\rho_a \min(Q_L, Q_a). \qquad (2.18)$$

The line-shape of the signal is determined by the kinetic energy distribution of the axion, which is typically assumed to be well approximated by the Maxwellian given by the isothermal sphere halo model [63]. However, further research has shown additional effects may result in modifications to the isothermal halo model [64–66]. There are also predictions of additional structure within the isothermal halo, which will be discussed further in Chap. 3 [67–72].

The parameters an experiment has control over is Q_L, B, and V. An experiment's sensitivity depends on how well these parameters are optimized. The frequency range of the cavity is determined by its size, which should be proportional to the axion's Compton wavelength. Thus, searching for lower mass axions with a longer Compton wavelength requires a very large cavity and higher mass axions require a smaller cavity.

2.3.2 ADMX

ADMX is a prime example of a Sikivie haloscope [73]. A diagram of the ADMX run 1c experimental set-up is shown in Fig. 2.11 and a diagram of the electronics in Fig. 2.12. A photo of the insert in the clean room is shown in Fig. 2.13. Two key technologies used by ADMX are a dilution refrigerator to reach mK temperatures and JPAs to significantly amplify the power of a signal [74, 75].

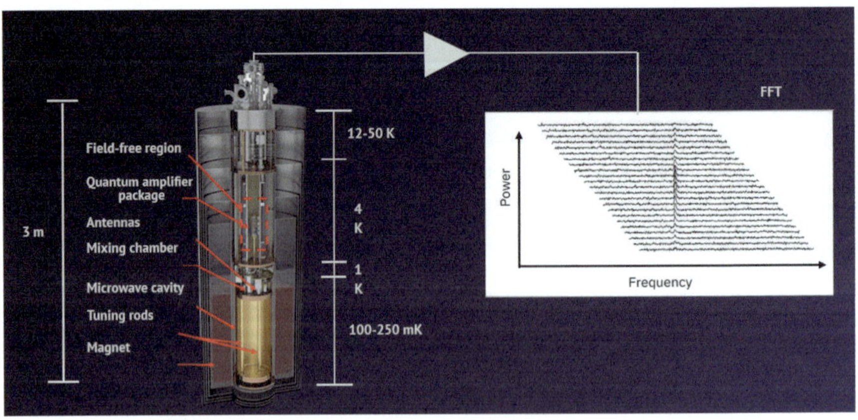

Fig. 2.11 Detailed diagram of the ADMX Run 1c set-up. The plot on the right is called a waterfall plot with subsequent power spectra lined up, showing what a signal from the isothermal halo model would look like

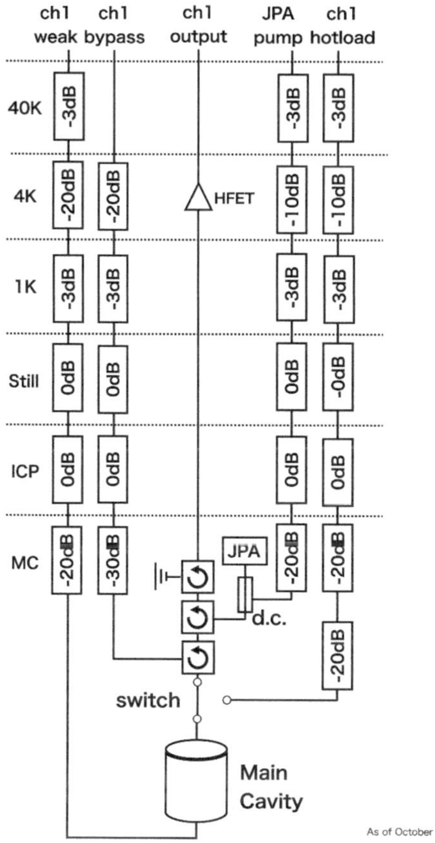

Fig. 2.12 Diagram of the electronics used in ADMX run 1c. On the left is the temperature scale, while each box contains the signal attenuation at each stage

2.3.3 Josephson Parametric Amplifiers

The dynamics of JPAs are similar to that of a child on a swing. For a child to increase their swinging height, they must pump their legs at the right frequency. For the JPA and signal from the cavity, the cavity signal is like the height of the child's swinging, the JPA acts as the swing, and a pump tone acts like the child pumping their legs. If the JPA is pumped with the correct frequency relative to the cavity frequency, energy can be transferred from the pump tone to the cavity signal [75]. The JPAs require a field-free region to operate optimally, shown in Fig. 2.11. The magnetic field of the superconducting magnet extends above the cavity, but is canceled out with a bucking coil magnet.

Fig. 2.13 A photo of the ADMX insert in the clean room. The height of the insert is approximately 3 m

2.3.4 *Cryogenics*

The experiment is cooled through the use of a dilution refrigerator. There are several temperature stages, beginning at room temperature at the top of the insert and dropping to approximately 100 mK at the cavity. There are many cables running from the warm space to the cavity, such as for temperature sensors and coaxial cables to send the cavity signal to a VNA. To prevent heat transfer through the cabling, they are heat sunk to each temperature stage.

Software has been written to control the cryogenic system remotely, which automates processes such as helium fills. ADMX also utilizes a liquifier, which significantly reduces the helium loss. The measurements of various sensors are plotted on an internal website, so the experiment can be monitored without needing to be physically in the lab. This ensures a fast response can be made in the event of unexpected behavior.

The cryogenic system is a vital element of the experiment, as thermal noise is the biggest contributor to the background signal. Reducing the system noise temperature to approximately 100 mK greatly increases the experiment's SNR.

2.3.5 *Signal-To-Noise Ratio*

The SNR is given by

$$\mathrm{SNR} = \frac{P_a}{k_B T_{sys}/\epsilon}\sqrt{\frac{t}{b}}, \tag{2.19}$$

where k_B is Boltzmann's constant, T_{sys} is the system noise temperature, ϵ is transmission efficiency between the cavity and JPA, t is the integration time, and b is the detection bandwidth. The SNR is estimated using a value called the SNR Improvement (SNRI), given by [75]

$$\mathrm{SNRI} = \frac{T_{sys}^{off}}{T_{sys}^{on}} = \frac{G_{on}}{G_{off}} \frac{P_{off}}{P_{on}}, \tag{2.20}$$

where $G_{on(off)}$ and $P_{on(off)}$ is the total gain and power spectral density of the rf chain with JPA on (off). When the JPA is turned off, the power is linearly proportional to the temperature, given by

$$P = G_{off} b k_B (T_{HL}\epsilon + T_{sys}^{off}), \tag{2.21}$$

where T_{HL} is the temperature of a hotload connected to the cavity, as shown in Fig. 2.12. This heat source can be varied between 0.5 K and 4 K. The power versus hotload temperature can plotted and fitted to determine the gain and system noise temperature with the JPA off. This procedure is referred to as a y-factor measurement [75]. Equation 2.20 can then be used to determine the system noise temperature with the JPA on. The average noise temperature of ADMX run 1c was approximately 600 mK.

2.3.6 ADMX Power and Scan Rate

Rewriting Eq. 2.18 such that the power of two experiments can be easily compared is more useful. Normalizing the expected axion power to the ADMX run 1c experimental parameters yields

$$\begin{aligned} \mathrm{P_{axion}} = 7.7 \times 10^{-23}\,\mathrm{W} \left(\frac{V}{136\,\ell}\right)\left(\frac{B}{7.5\,\mathrm{T}}\right)^2\left(\frac{C}{0.4}\right)\left(\frac{g_\gamma}{0.36}\right)^2 \\ \times \left(\frac{\rho_a}{0.45\,\mathrm{GeV/cc}}\right)\left(\frac{f}{1\,\mathrm{GHz}}\right)\left(\frac{Q_L}{80,000}\right), \end{aligned} \tag{2.22}$$

where g_γ is defined by Eq. A.6 and C is the form factor for the TM_{010} mode, calculated using Eq. 2.16.

The scan rate of the experiment can be determined by substituting the expected axion power into Eq. 2.19 and computing df/dt, which yields

$$\frac{df}{dt} \approx 277 \frac{\text{MHz}}{\text{yr}} \left(\frac{V}{136\,\ell}\right)^2 \left(\frac{B}{7.5\,\text{T}}\right)^4 \left(\frac{C}{0.4}\right)^2 \left(\frac{g_\gamma}{0.36}\right)^4 \left(\frac{3.5}{\text{SNR}}\right)^2$$
$$\times \left(\frac{\rho_a}{0.45\ \text{GeV/cc}}\right)^2 \left(\frac{f}{1\ \text{GHz}}\right)^2 \left(\frac{Q_L}{80,000}\right) \left(\frac{0.6\ \text{K}}{T_{sys}}\right)^2. \quad (2.23)$$

2.3.7 ADMX Current Limits

ADMX run 1a, 1b, and 1c excluded the KSVZ axion across the mass range of approximately 2.66 μeV to 4.2 μeV, as well as the DFSZ axion in a large fraction of the range [74–76]. ADMX was the first experiment to reach DFSZ sensitivity. Unless a new, well-motivated theory predicts an axion with a coupling even weaker than that of DFSZ, these limits completely rule out the axion in this mass range. The limits are shown in Fig. 2.14.

2.3.8 Other Axion Experiments

There are many other searches for axions. Experiments such as HAYSTAC, CAPP, and ORGAN are traditional cavity haloscopes, searching for higher mass axions, above 10 μeV [77–79]. There also lumped-element circuit resonator searches, primarily used to search for low mass axions, like ADMX SLIC and DM Radio [80–83].

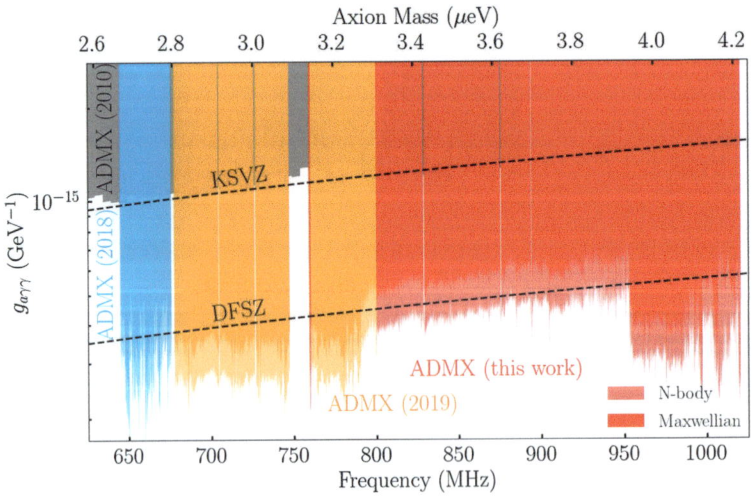

Fig. 2.14 ADMX exclusion plot for runs 1a, 1b, and 1c, covering approximately 650 MHz to 1020 MHz

Another is the CERN Axion Solar Telescope (CAST), which aims to detect axions emitted by the sun [84]. To name one more type of search, there are proposals to search for axions using superfluid helium-3 and helium-3 gas [85, 86].

This list surely does not include every experiment currently being run, prototyped, or proposed, as the there has been an explosion of ideas for experiments over the last several years. The search for axions has become incredibly diverse, utilizing experimental techniques from a broad range of fields.

References

1. S. Weinberg, Phys. Rev. Lett. **40**, 223 (1978). https://doi.org/10.1103/PhysRevLett.40.223
2. B. Beltrán, M. Kuster, G.G. Raffelt, *Axions: Theory, Cosmology, and Experimental Searches*. Lecture Notes in Physics No. 741 (Springer, Berlin New York, 2008)
3. M.D. Schwartz, *Quantum Field Theory and the Standard Model*, 1st edn. (Cambridge University Press, 2013). https://doi.org/10.1017/9781139540940
4. S. Weinberg, Phys. Rev. D **11**, 3583 (1975). https://doi.org/10.1103/PhysRevD.11.3583
5. G. Hooft, Phys. Rep. **142**, 357 (1986). https://doi.org/10.1016/0370-1573(86)90117-1
6. G. Hooft, Phys. Rev. Lett. **37**, 8 (1976). https://doi.org/10.1103/PhysRevLett.37.8
7. S.L. Adler, Phys. Rev. **177**, 2426 (1969). https://doi.org/10.1103/PhysRev.177.2426
8. J.S. Bell, R. Jackiw, Nuov Cim. A **60**, 47 (1969). https://doi.org/10.1007/BF02823296
9. C.S. Wu et al., Phys. Rev. **105**, 1413 (1957). https://doi.org/10.1103/PhysRev.105.1413
10. J.H. Christenson et al., Phys. Rev. Lett. **13**, 138 (1964). https://doi.org/10.1103/PhysRevLett.13.138
11. B. Aubert et al., Phys. Rev. Lett. **87**, 091801 (2001). https://doi.org/10.1103/PhysRevLett.87.091801
12. R. Aaij et al., J. High Energy Phys. **2014**, 41 (2014). https://doi.org/10.1007/JHEP07(2014)041
13. V. Baluni, Phys. Rev. D **19**, 2227 (1979). https://doi.org/10.1103/PhysRevD.19.2227
14. R. Crewther et al., Phys. Lett. B **88**, 123 (1979). https://doi.org/10.1016/0370-2693(79)90128-X
15. J.M. Pendlebury et al., Phys. Rev. D **92**, 092003 (2015). https://doi.org/10.1103/PhysRevD.92.092003
16. J. Gasser, H. Leutwyler, Phys. Rep. **87**, 77 (1982). https://doi.org/10.1016/0370-1573(82)90035-7
17. R.D. Peccei, H.R. Quinn, Phys. Rev. Lett. **38**, 1440 (1977). https://doi.org/10.1103/PhysRevLett.38.1440
18. F. Wilczek, Phys. Rev. Lett. **40**, 279 (1978). https://doi.org/10.1103/PhysRevLett.40.279
19. H. Faissner et al., Phys. Lett. B **96**, 201 (1980). https://doi.org/10.1016/0370-2693(80)90244-0
20. J.E. Kim, Phys. Rep. **150**, 1 (1987). https://doi.org/10.1016/0370-1573(87)90017-2
21. J.E. Kim, Phys. Rev. Lett. **43**, 103 (1979). https://doi.org/10.1103/PhysRevLett.43.103
22. M. Shifman, A. Vainshtein, V. Zakharov, Nucl. Phys. B **166**, 493 (1980). https://doi.org/10.1016/0550-3213(80)90209-6
23. M. Dine, W. Fischler, M. Srednicki, Phys. Lett. B **104**, 199 (1981). https://doi.org/10.1016/0370-2693(81)90590-6
24. A.R. Zhitnitsky, Sov. J. Nucl. Phys. **31**, 260 (1980)
25. P. Svrcek, E. Witten, J. High Energy Phys. **2006**, 051 (2006). https://doi.org/10.1088/1126-6708/2006/06/051
26. A. Ernst, L. Di Luzio, A. Ringwald, C. Tamarit, *Proceedings of Corfu Summer Institute 2018 "School and Workshops on Elementary Particle Physics and Gravity"—PoS (CORFU2018)* (Sissa Medialab, Corfu, Greece, 2019), p. 054. https://doi.org/10.22323/1.347.0054

27. J. Engel, D. Seckel, A.C. Hayes, Phys. Rev. Lett. **65**, 960 (1990). https://doi.org/10.1103/PhysRevLett.65.960
28. N. Iwamoto, Phys. Rev. Lett. **53**, 1198 (1984). https://doi.org/10.1103/PhysRevLett.53.1198
29. A. Caputo, G. Raffelt, Astrophysical axion bounds: the 2024 edition (2024). Preprint at arXiv:2401.13728
30. A. Araujo, D.F. López, J.G. Pereira, Gravit. Cosmol. **25**, 157 (2019). https://doi.org/10.1134/S0202289319020026
31. F. Zwicky, ApJ **86**, 217 (1937). https://doi.org/10.1086/143864
32. J.S. Lewis, *Physics and Chemistry of the Solar System* (Academic Press, San Diego, 1995)
33. V.C. Rubin, N. Thonnard, W.K. Ford Jr., ApJ **238**, 471 (1980). https://doi.org/10.1086/158003
34. D. Clowe et al., ApJ **648**, L109 (2006). https://doi.org/10.1086/508162
35. M. Markevitch, A.H. Gonzalez, D. Clowe, A. Vikhlinin, W. Forman, C. Jones, S. Murray, W. Tucker, ApJ **606**, 819 (2004). https://doi.org/10.1086/383178
36. S. Weinberg, *Cosmology* (Oxford University Press, New York, 2008)
37. D. Baumann, *Cosmology* (Cambridge University Press, United Kingdom, New York Port, Melbourne New Delhi Singapore, Cambridge, 2022)
38. N. Aghanim, Y. Akrami, et al., A and A **641** (2020). https://doi.org/10.1051/0004-6361/201833910e
39. S. Schindler (2001). arXiv:arXiv:0107028
40. G. Bertone, D. Hooper, Rev. Mod. Phys **90**, 045002 (2018). https://doi.org/10.1103/RevModPhys.90.045002
41. S. Dodelson, L.M. Widrow, Phys. Rev. Lett. **72**, 17 (1994). https://doi.org/10.1103/PhysRevLett.72.17
42. S. Giagu, Front. Phys. **7**, 75 (2019). https://doi.org/10.3389/fphy.2019.00075
43. M. Selvi, Xenon Collaboration, *Proceedings of Neutrino Oscillation Workshop—PoS(NOW2018)* (Sissa Medialab, Rosa Marina (Ostuni, Brindisi, Italy), 2019), p. 094. https://doi.org/10.22323/1.337.0094
44. D.N. McKinsey, LZ Collaboration, J. Phys. Conf. Ser **718**, 042039 (2016). https://doi.org/10.1088/1742-6596/718/4/042039
45. C.E. Aalseth et al., Eur. Phys. J. Plus **133**, 131 (2018). https://doi.org/10.1140/epjp/i2018-11973-4
46. R. Agnese, SuperCDMS Collaboration, Phys. Rev. Lett. **112**, 241302 (2014). https://doi.org/10.1103/PhysRevLett.112.241302
47. J. Aalbers et al., First dark matter search results from the LUX-ZEPLIN (LZ) experiment (2022). Preprint at arXiv:2207.03764 [astro-ph, physics:hep-ex]
48. C.A.J. O'Hare, Phys. Rev. Lett. **127**, 251802 (2021). https://doi.org/10.1103/PhysRevLett.127.251802
49. R.H. Sanders, S.S. McGaugh, Annu. Rev. Astron. Astrophys. **40**, 263 (2002). https://doi.org/10.1146/annurev.astro.40.060401.093923
50. C. Skordis, T. Zlosnik, Phys. Rev. Lett. **127**, 161302 (2021). https://doi.org/10.1103/PhysRevLett.127.161302
51. M. Milgrom, ApJ **270**, 365 (1983). https://doi.org/10.1086/161130
52. P.M. Chesler, A. Loeb, Phys. Rev. Lett. **119**, 031102 (2017). https://doi.org/10.1103/PhysRevLett.119.031102
53. V. Sahni, *The Physics of the Early Universe*, vol. 653, ed. by E. Papantonopoulos (Springer Berlin Heidelberg, Berlin, Heidelberg, 2004), pp. 141–179. Lecture Notes in Physics. https://doi.org/10.1007/978-3-540-31535-3_5
54. E. Verlinde, SciPost Phys. **2**, 016 (2017). https://doi.org/10.21468/SciPostPhys.2.3.016
55. P. Sikivie, Phys. Rev. Lett. **48**, 1156 (1982). https://doi.org/10.1103/PhysRevLett.48.1156
56. E. Witten, Ann. Phys. **128**, 363 (1980). https://doi.org/10.1016/0003-4916(80)90325-5
57. S. Chang, C. Hagmann, P. Sikivie, Phys. Rev. D **59**, 023505 (1998). https://doi.org/10.1103/PhysRevD.59.023505
58. J. Preskill, M.B. Wise, F. Wilczek, Phys. Lett. B **120**, 127 (1983). https://doi.org/10.1016/0370-2693(83)90637-8

59. L. Abbott, P. Sikivie, Phys. Lett. B **120**, 133 (1983). https://doi.org/10.1016/0370-2693(83)90638-X
60. J. Ipser, P. Sikivie, Phys. Rev. Lett **50**, 925 (1983). https://doi.org/10.1103/PhysRevLett.50.925
61. P. Sikivie, Phys. Rev. Lett **51**, 1415 (1983). https://doi.org/10.1103/PhysRevLett.51.1415
62. P. Sikivie, Phys. Rev. D **32**, 2988 (1985). https://doi.org/10.1103/PhysRevD.32.2988
63. M.S. Turner, Phys. Rev. D **33**, 889 (1986). https://doi.org/10.1103/PhysRevD.33.889
64. J.F. Navarro, C.S. Frenk, S.D.M. White, ApJ **462**, 563 (1996). https://doi.org/10.1086/177173
65. E.W. Lentz, T.R. Quinn, L.J. Rosenberg, M.J. Tremmel, ApJ **845**, 121 (2017). https://doi.org/10.3847/1538-4357/aa80dd
66. P. Sikivie, J.R. Ipser, Phys. Lett. B **291**, 288 (1992). https://doi.org/10.1016/0370-2693(92)91047-D
67. P. Sikivie, Phys. Lett. B **432**, 139 (1998). https://doi.org/10.1016/S0370-2693(98)00595-4
68. P. Sikivie, Phys. Rev. D **60**, 063501 (1999). https://doi.org/10.1103/PhysRevD.60.063501
69. L.D. Duffy, P. Sikivie, Phys. Rev. D **78**, 063508 (2008). https://doi.org/10.1103/PhysRevD.78.063508
70. L.D. Duffy et al., Phys. Rev. D **74**, 012006 (2006). https://doi.org/10.1103/PhysRevD.74.012006
71. J.D. Hoskins, A modulation sensitive search for non-virialized axions in the Milky Way halo, Ph.D. thesis, school University of Florida, 2014
72. J. Hoskins et al., Phys. Rev. D **94**, 082001 (2016). https://doi.org/10.1103/PhysRevD.94.082001
73. L.J. Rosenberg, Proc. Natl. Acad. Sci. U.S.A **112**, 12278 (2015). https://doi.org/10.1073/pnas.1308788112
74. C. Bartram, ADMX Collaboration, Phys. Rev. D **103**, 032002 (2021). https://doi.org/10.1103/PhysRevD.103.032002
75. C. Bartram et al., Phys. Rev. Lett. **127**, 261803 (2021). https://doi.org/10.1103/PhysRevLett.127.261803
76. C. Boutan et al., Axion dark matter experiment: Run 1A analysis details (2023). Preprint at arXiv:2312.16668
77. L. Zhong et al., Phys. Rev. D **97**, 092001 (2018). https://doi.org/10.1103/PhysRevD.97.092001
78. O. Kwon et al., Phys. Rev. Lett. **126**, 191802 (2021). https://doi.org/10.1103/PhysRevLett.126.191802
79. A. Quiskamp, B.T. McAllister, P. Altin, E.N. Ivanov, M. Goryachev, M.E. Tobar, Phys. Rev. Lett. **132**, 031601 (2024). https://doi.org/10.1103/PhysRevLett.132.031601
80. N. Crisosto et al., Phys. Rev. Lett. **124**, 241101 (2020). https://doi.org/10.1103/PhysRevLett.124.241101
81. P. Sikivie, N. Sullivan, D. Tanner, Phys. Rev. Lett. **112**, 131301 (2014). https://doi.org/10.1103/PhysRevLett.112.131301
82. S. Chaudhuri, P.W. Graham, K. Irwin, J. Mardon, S. Rajendran, Y. Zhao, Phys. Rev. D **92**, 075012 (2015). https://doi.org/10.1103/PhysRevD.92.075012
83. L. Brouwer, DMRadio Collaboration, Phys. Rev. D **106**, 112003 (2022). https://doi.org/10.1103/PhysRevD.106.112003
84. CAST Collaboration, Nat. Phys. **13**, 584 (2017). https://doi.org/10.1038/nphys4109
85. C. Gao, W. Halperin, Y. Kahn, M. Nguyen, J. Schütte-Engel, J.W. Scott, Phys. Rev. Lett. **129**, 211801 (2022). https://doi.org/10.1103/PhysRevLett.129.211801
86. A.A. Geraci for the Ariadne Collaboration et al., *Microwave Cavities and Detectors for Axion Research*, vol. 211, ed. by G. Carosi, G. Rybka, K. Van Bibber (Springer International Publishing, Cham, 2018), pp. 151–161. Springer Proceedings in Physics. https://doi.org/10.1007/978-3-319-92726-8_18

Chapter 3
ADMX Multi-Resolution Analysis

Abstract Many axion dark matter experiments search for an axion signal with a Maxwell Boltzmann distribution, predicted by the isothermal halo model. However, recent work has shown that the dark matter distribution may be more exotic, such as containing caustics or being entirely composed of miniclusters. A key feature of these models is the signal becomes much narrower. This chapter introduces the effect of narrow signals on detection and describes the analysis of ADMX run 1c data, placing limits on exotic halo structure. Ultimately, this analysis rules out narrow halo structures with $\rho_a \gtrsim 0.03\,\mathrm{GeV/cm^3}$, over the mass range $3.3 - 4.2\,\mu$eV.

3.1 Motivation

Searches for axion dark matter typically assume the isothermal halo model. This model produces an axion signal with a Maxwell-Boltzmann line shape in frequency space. The caustic ring model predicts a small fraction of the dark matter within the halo would be present in cold, nonvirialized axion flows [1–3]. Several searches for these features using ADMX data have been completed [4–8]. The key feature of these flows are their signals being especially sharp in the frequency domain.

Figure 3.1 depicts a cartoon of cold, nonvirialized axion signals overlaid the isothermal halo signal. The pairs of signals correspond to flows of the same order, for example axions that are falling in for the first time and falling out for the first time form a pair, and so on for the continued motion.

The width of an axion signal can be approximated by assuming values for its frequency, velocity, and velocity dispersion. The total energy of an axion, E_a, can be written as

$$\hbar\omega = \frac{m_a c^2}{\sqrt{1-\frac{v^2}{c^2}}} = m_a c^2 \left[1 + \frac{1}{2}\frac{v^2}{c^2} + O(v^4)\right], \tag{3.1}$$

A. Hipp, *Searching High and Low for the Axion*,
Springer Theses, https://doi.org/10.1007/978-3-032-02276-9_3

Fig. 3.1 Cartoon of cold, nonvirialized axion signals

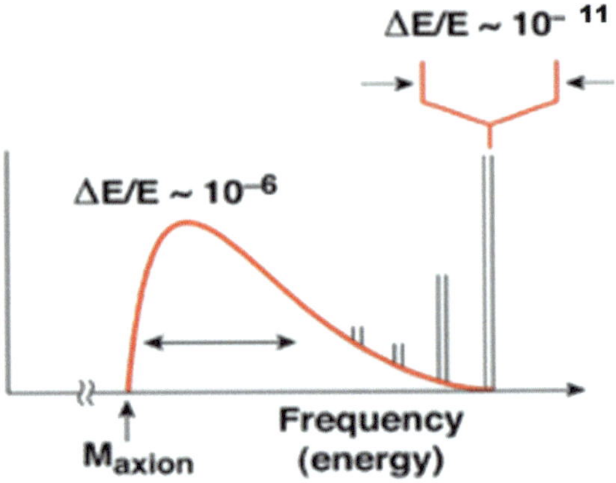

where $\hbar$ and c are Planck's constant and the speed of light, respectively, m_a is the mass of the axion, and v is the mean relative velocity of the axion flow. The higher order terms can be neglected. The differential is given by

$$\hbar\delta\omega = m_a v\delta v. \tag{3.2}$$

The central frequency of the axion is given by $\hbar\omega = m_a c^2$. Dividing Eq. 3.2 by this value and substituting in f for ω yields

$$\frac{\delta f}{f} = \frac{v\delta v}{c^2}. \tag{3.3}$$

Equation 3.3 yields an approximate frequency width. In the isothermal halo model, the axion velocity is assumed to be in the range of 0 to 300 km/s, which makes the velocity dispersion on the order of $O(10^{-3}c)$. At a frequency of 1 GHz, this produces a frequency width of about 1 kHz. In the caustic ring model, one well studied flow is referred to as the big flow [6, 9]. The big flow is estimated to have a velocity of 480 km/s and a velocity dispersion of 53 m/s. At a frequency of 1 GHz, these values yield a frequency width of approximately 283 mHz. Unless both signals have an incredible amount of power, each requires a different frequency resolution to be detected.

A feature that would require a high frequency resolution to detect is the modulation of an axion signal due to the motion of the Earth [10, 11]. As the Earth rotates around its axis and orbits the sun, the frequency of an axion signal will be modulated. The maximum modulation over the course of a day is 1 Hz, and over one year 100 Hz.

ADMX has two analysis channels to accommodate these features. The Medium Resolution (MedRes) channel has a frequency resolution 200 Hz and is the main analysis channel. Second is the High Resolution (HiRes) channel. HiRes begins their analysis with the time series data, allowing for power spectra of varying resolutions to be computed, with a resolution of 20 mHz being the finest that can be computed.

3.2 Producing Power Spectra

3.2.1 ADMX DAQ

The AMDX DAQ produces the MedRes and HiRes data sets in parallel. It has a sampling rate of 50 MS/s and a buffer of $n = 500,000$ points. Each scan is 100 s of data taking. Every 10 ms throughout the 100 s scan, a Fast Fourier Transform (FFT) is computed producing a symmetric amplitude spectrum. Only half of the spectrum, along with the center bin, is kept to have an asymmetric, 250,001-point amplitude spectrum. Next, 500 bins around the center frequency are chosen, corresponding to a frequency range of 50 kHz. This is typically referred to as the Fourier frequency range, spanning 0 to 50 kHz, where 25 kHz corresponds to the resonant frequency of the cavity. At this point, the process diverges for MedRes and HiRes.

For MedRes, a running average is taken; the power spectra are computed, multiplied by a normalization factor, and summed. The final power spectrum is the average of 10,000 sub-spectra.

For HiRes, the inverse FFT of the 500-point amplitude spectra are taken and multiplied by a scaling factor, producing a complex time-series. The 10,000 complex time series are concatenated forming a single 5,000,000-point complex time series, spanning 100 s. The flow chart below outlines this process.

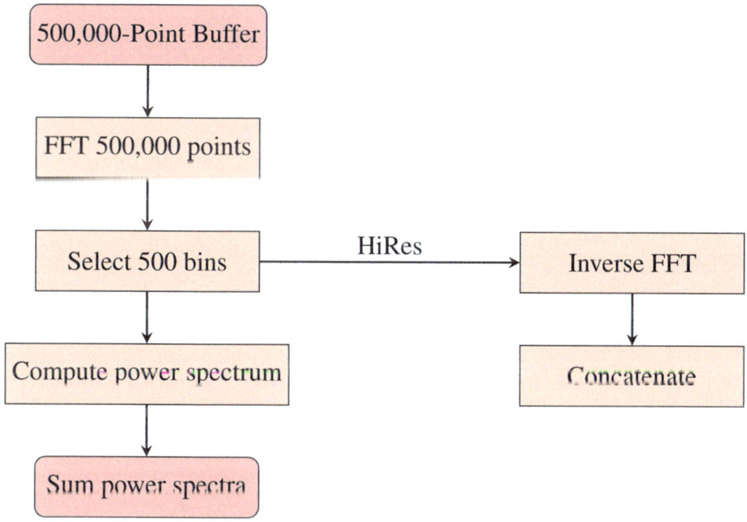

The time-series is complex because the inverse Fourier transform to go back to the time domain is performed on an asymmetric amplitude spectrum. Histograms of the complex phase for a time-series and its associated amplitude spectrum are shown in Fig. 3.2.

The complex phase follows a uniform distribution, confirming the inverse FFT introduced an arbitrary phase.

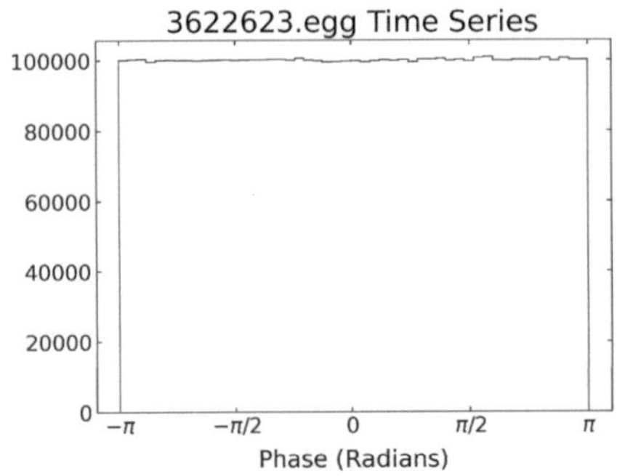

(a) Time Series.

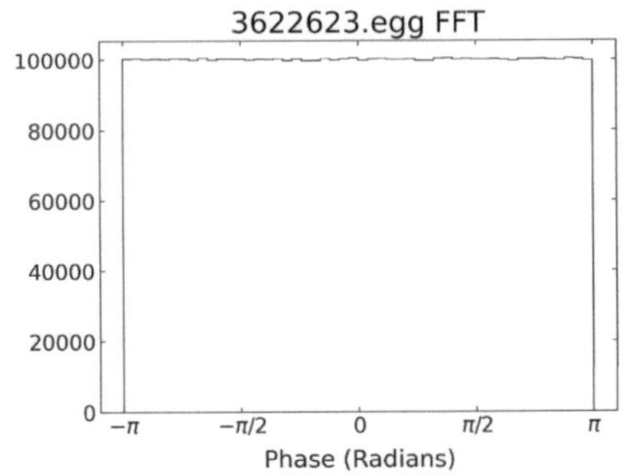

(b) Amplitude Spectrum.

Fig. 3.2 Histograms of the complex phase

3.2.2 Computing Normalized Spectra

Starting with the complex time series, power spectra with varying frequency resolutions can be computed by dividing the time series into equal length arrays. In theory, a large range of resolutions could be chosen, but in practice only resolutions including all 5,000,000 points are used. The power built up by a signal depends on the integration time, so the full 100 s of data should be kept.

The frequency resolution for a spectrum with n points is given by

$$f_{resolution} = 2 * \frac{50,000}{n} \mathrm{Hz}, \tag{3.4}$$

where the factor of 2 corresponds to peaks needing to be separated by at least one bin to be properly resolved. With the desired resolution chosen, the complex time series is divided into $\mathrm{N} = \frac{5,000,000}{n}$ n-point sub-series. The power spectra are computed by squaring the absolute value of the amplitude spectra. The absolute value must be taken first as the FFT of a complex time series produces a complex amplitude spectrum.

For $\mathrm{N} = 1$, there is a single 5,000,000-point power spectrum, with a frequency resolution of 20 mHz.

Figure 3.3 shows an example unnormalized power spectrum. The slight "U" shape of spectrum is the characteristic receiver shape. This shape is removed by normalizing the power spectrum with a polynomial fit. Although normalizing a single 5,000,000-point spectrum is not computationally intensive, doing so for 100,000 scans is very time consuming. Decimating the spectrum prior to normalization eases the computational load. In practice, the power spectrum is decimated by a factor of 20, choosing only every 20th point, and the 250,000-point spectrum is fit with a 6th degree polynomial. The unnormalized power spectrum is divided by this fit yielding the desired normalized power spectrum.

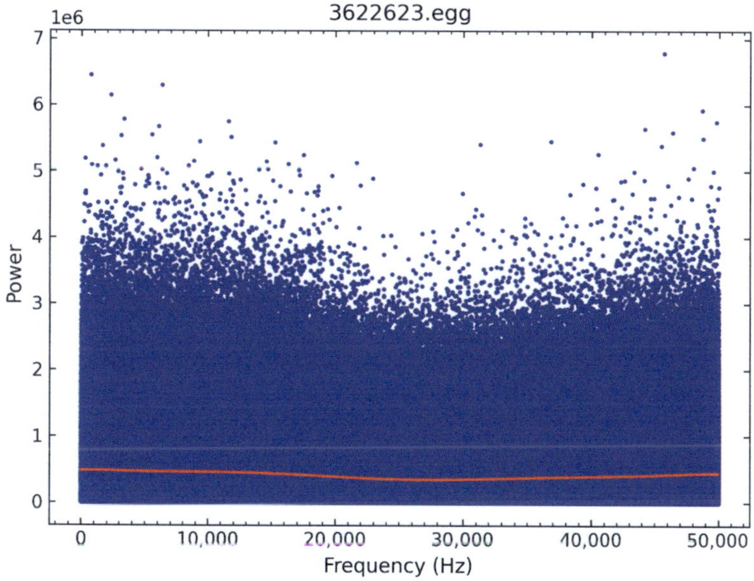

Fig. 3.3 An example unnormalized power spectrum.The red line is the polynomial fit

For $N \geq 2$, the correct statistics are achieved only when independent, identically distributed random variables are added. The statistics will be discussed more in Sect. 3.2.3. The independent condition is automatically satisfied for the N sub-spectra. After normalizing the sub-spectra, they are nearly identically distributed, satisfying the second condition. Fitting all sub-spectra separately is again computationally intensive, so a different approach is utilized. It is reasonable to suspect the receiver shape stays nearly constant during the 100 s. If this is the case, simply fitting the first sub-spectrum and using the resulting polynomial to normalize all sub-spectra is sufficient.

A test was performed on a sample of 1,465 digitizations, comparing the mean and standard deviation of the power spectra computed after normalizing all sub-spectra independently versus only fitting the first. This was done for resolutions of 200 mHz and 1000 mHz, corresponding to the sum of 10 and 50 sub-spectra, respectively. The sample included files with only background, with synthetic axion signals, and with RFI to compare performance in all expected circumstances. For 200 mHz, the average difference between the two means and standard deviations was 0.03σ and 0.25σ, respectively, where these values are in units of normalized power. This resolution has a typical mean of 20σ and standard deviation of $\sqrt{20}\sigma$, so the difference is negligible.

For 1000 mHz, the average difference between the two means and standard deviations was 0.13σ and 0.75σ, respectively. This resolution has a typical mean of 50σ and standard deviation of $\sqrt{50}\sigma$, so again the difference is negligible. The worst case scenario for 1000 mHz was a difference of 1σ between the means for a digitization with a strong synthetic signal. This difference did not prevent the signal from being discernible. Thus, this test showed only normalizing the first sub-spectrum is sufficient.

3.2.3 *Statistics of Power Spectra*

In general, the power spectrum follows an Erlang distribution [4]. The Erlang distribution is the sum of independent, identically distributed exponential distributions and the general form is given by

$$f(x; k, \lambda) = \frac{\lambda^k x^{k-1} e^{-\lambda x}}{(k-1)!}, \tag{3.5}$$

where $x, \lambda \in [0, \infty)$ and k is any positive integer. In this analysis, x corresponds to the power in units of normalized power, k is the number of sub-spectra being summed, and λ is the reciprocal of the variance of the power. For the sum of n sub-spectra, Eq. 3.5 can be rewritten to yield Eq. 3.6

$$\frac{dP}{dp_n} = \frac{p_n^{n-1}}{(n-1)!\sigma_a^{2n}} e^{\frac{p_n}{\sigma_a^{2n}}}, \tag{3.6}$$

where $p \in [0, \infty)$ is the power and σ_a is the standard deviation of the identically distributed sub-spectra. Normalizing the sub-spectra prior to summing them forces $\sigma_a = 1$ and Eq. 3.6 simplifies to

$$\frac{dP}{dp_n} = \frac{p_n^{n-1}}{(n-1)!} e^{p_n}. \tag{3.7}$$

For analyzing the ADMX run 1c data, frequency resolutions of 20 mHz, 40 mHz, 200 mHz, and 1000 mHz were chosen, corresponding to the sum of 1, 2, 10, and 50 sub-spectra, respectively. For 20 mHz, the normalized power spectrum has a mean and standard deviation of 1σ, which is the reason for the units of normalized power being labeled by σ. Figures 3.4, 3.5, 3.6, and 3.7 are histograms of the power for a digitization at each resolution.

As expected from the central limit theorem, as n increases, the distribution approaches a Gaussian distribution. Equation 3.7 was used to determine suitable power thresholds, which will be discussed in the coming section.

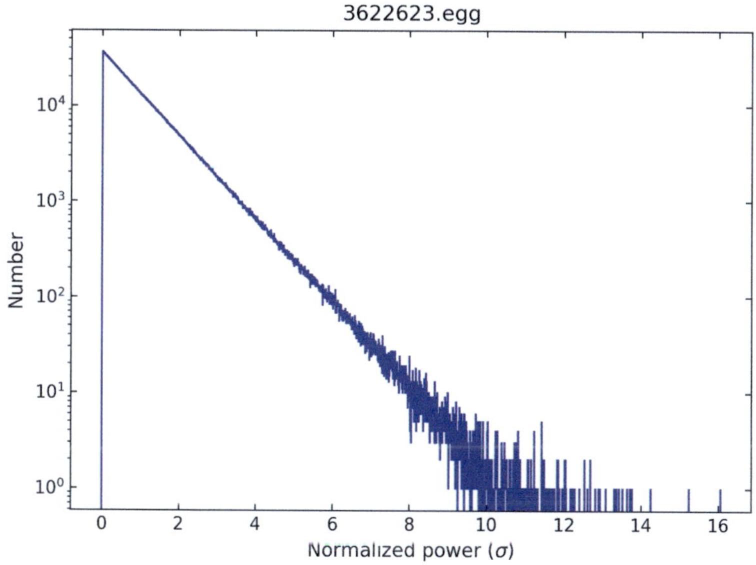

Fig. 3.4 Histogram of a power spectrum with a frequency resolution of 20 mHz

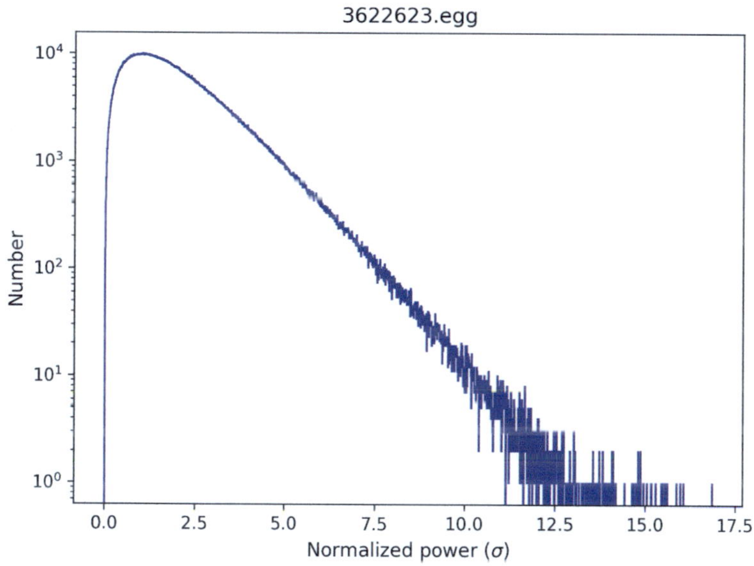

Fig. 3.5 Histogram of a power spectrum with a frequency resolution of 40 mHz

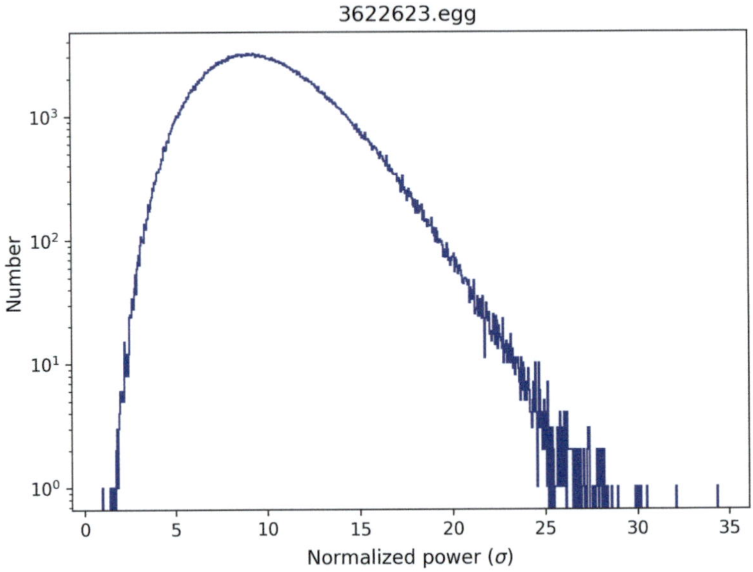

Fig. 3.6 Histogram of a power spectrum with a frequency resolution of 200 mHz

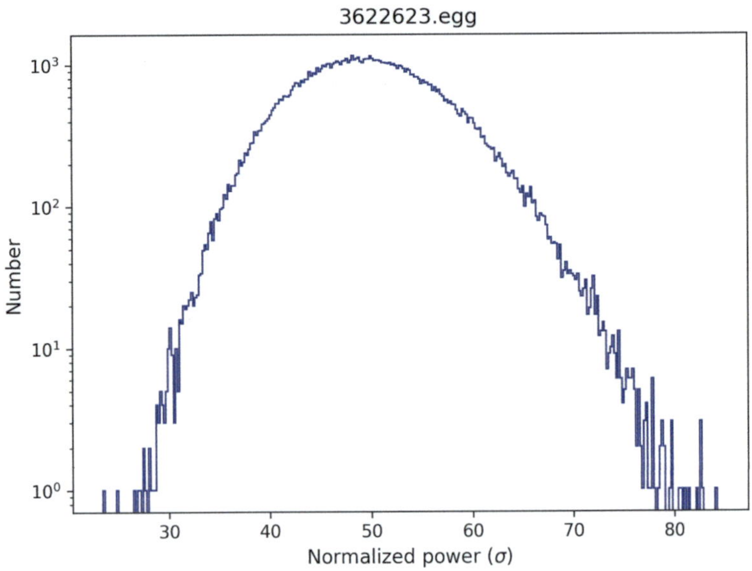

Fig. 3.7 Histogram of a power spectrum with a frequency resolution of 1000 mHz

3.2.4 MultiRes Motivation

As previously mentioned, a single bin search was performed at frequency resolutions of 20, 40, 200, and 1000 mHz. The bin width is half the frequency resolution. The signal-to-noise ratio (SNR) of a signal depends on the width of the signal relative to the bin width. Consider an example.

Suppose there is a signal with a FWHM 1 Hz. At the finest resolution of 20 mHz, the power is divided amongst 100 bins. The total power of the signal is conserved at all resolutions, so in this case each bin gets only a small fraction of the power. If the signal is relatively weak, this division will result in the power being below the noise floor. On the contrary, if the frequency resolution 200 Hz, like MedRes, the signal is will be averaged with 99 other bins. Again, unless the power is sufficiently large the signal will be below the noise floor. The signal will only become visible when the frequency resolution is adjusted properly.

One definition for SNR is given by:

$$\text{SNR} = \frac{\mu}{\sigma}, \tag{3.8}$$

where μ is the mean power of the signal and σ is the standard deviation of the noise background.

A study was performed on a simulated signal and the SNR was calculated at various resolutions to determine the relationship. First, the time series of the signal was computed, then the power spectrum, then finally the SNR was calculated. In the simulation, an Erlang distribution was assumed for the background, as is the case in ADMX. The simulated signal was built up from decaying exponential time series, which yields a approximately Lorentzian line shape in frequency space.

The time series was computed using

$$A(t) = e^{-T},$$

where T is a 5,000,000-point set of times spanning from 0 to 100 s. This produced a power spectrum spanning −25 to +25 kHz. The resulting power spectra for frequency resolutions of 20, 80, and 160 mHz are shown in Fig. 3.8a–c.

Since the HiRes search is a single bin search, a modified SNR was computed using the maximum power of the signal, rather than the mean power. This yielded a more applicable result. The SNR versus frequency resolution is shown in Fig. 3.9.

The maximum SNR was for a frequency resolution of 800 mHz, corresponding to a bin width of 400 mHz. The Lorentzian like signal had a FWHM of approximately 320 mHz. Thus, a bin width just larger than the FWHM gave the best SNR, which is an intuitive result.

The above example was for a symmetric signal, but what if that is not the case? Another simulated signal was computed, again using a decaying exponential signal in time, spanning 100 s, but now was multiplied by a sum of sine waves. This yielded Lorentzian-like signals, mixed to the frequency of the sine waves. If the frequency

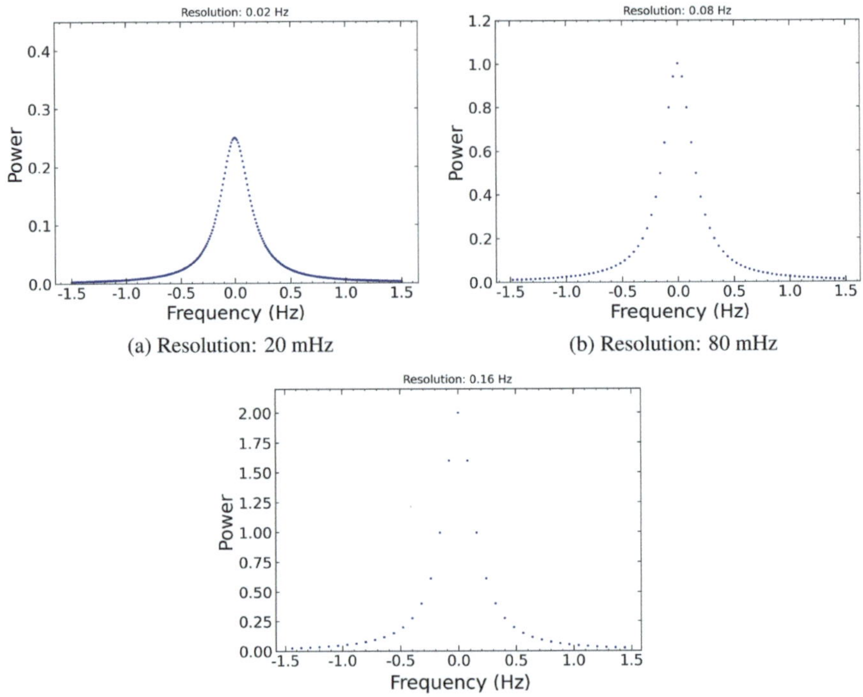

(a) Resolution: 20 mHz

(b) Resolution: 80 mHz

(c) Resolution: 160 mHz

Fig. 3.8 Lorentzian like signal produced using three frequency resolutions

difference of the sine waves was smaller than the bin width, it appeared as a single, continuous signal. The amplitude of each sine wave was modulated to produce a power spectrum with a Maxwell-Boltzmann-like line-shape. The time series signal was given by

$$A(t) = e^{-t} \times \sum_{n=1}^{N} b\frac{n^2}{a} e^{\frac{-n^2}{2(a/4)^2}} sin(2(f_0 + n\Delta f)\pi t), \tag{3.9}$$

where N was the number of sine waves being summed, the coefficient $b\frac{n^2}{a^3}e^{-n/(2a^2)}$ yielded an approximate Maxwell-Boltzmann line-shape with a and b controlling the width and shape, f_0 was the start frequency of the signal, and Δf was the frequency spacing between sine waves.

Values of $N = 1000$, $f_0 = 10,000$ Hz, $\Delta f = 0.05$ Hz, $b = 64$, and $a = 1000$ were used. The resulting signal at a frequency resolution of 20 mHz is shown in Fig. 3.10.

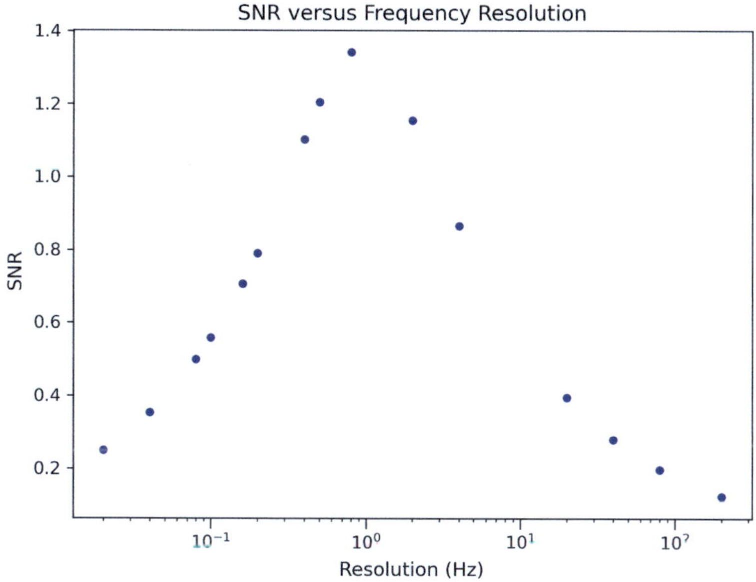

Fig. 3.9 SNR versus frequency resolution for a Lorentzian-like signal

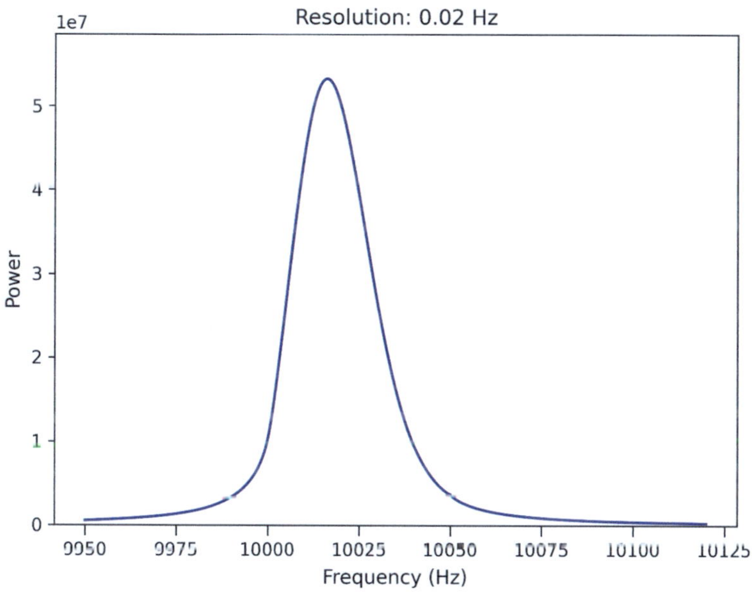

Fig. 3.10 Maxwell-Boltzmann like signal produced using the described process at a frequency resolution of 20 mHz

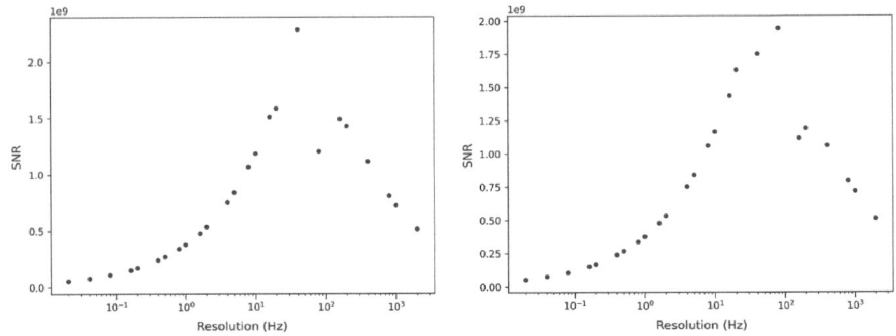

(a) Start frequency of 10,000 Hz (b) start frequency of 10,011 Hz

Fig. 3.11 Comparison of SNR versus frequency resolution for the same signal, with a shifted resonant frequency

The approximate FWHM of this signal 25 Hz. The SNR versus frequency resolution is shown in Fig. 3.11a. Changing the start frequency to $f_0 = 10,011$, the SNR versus resolution in Fig. 3.11b was found.

The resolution at which the maximum SNR is achieved changes 40 Hz to the next largest computed 80 Hz. The difference is due to the signal shifting with respect to the center of the nearest bin, so the amount of power deposited into the nearest bin decreased. However, the difference in SNR was quite small, suggesting this to be a non-issue. The main finding is that again the maximum SNR was achieved at approximately the FWHM of the signal.

Figures 3.11 and 3.9 show the SNR of a signal changes significantly with frequency resolution. Choosing the correct frequency resolution is important; however, that is quite difficult. The flows in the caustic ring model have varying signal widths, and being agnostic of what could be detected, there is no way to know the best frequency resolution. Thus, analyzing the data at various frequency resolutions should be done.

3.3 Analysis of ADMX Run 1c Data

A single bin search was performed on the entire run 1c part 1 data set, which spanned an axion mass range of approximately 3.3 to 4.2 μeV. This corresponded to recording all individual frequency bins with excess power. The data set consisted of 132,992 good digitizations, which refers to digitizations recorded during data taking times with an integration time of 100 s. Significantly more data than this was recorded but for the purpose of calibration and troubleshooting and thus not considered quality data.

3.3.1 Producing the Trigger List

Excess power was determined using Eq. 3.7. Bins with excess power are referred to as triggers. First, a power threshold for a frequency resolution of 20 mHz was found. This was done by performing a linear fit of the histogram of power, calculating where the line intersected $y = 1$, then adding three to this value. This yielded a cutoff of 13.9 σ, which gives an expectation of

$$5,000,000 * \int_{13.9}^{\infty} e^{-p} dp = 4.59, \tag{3.10}$$

triggers per digitization. The trigger cutoffs for the remaining frequency resolutions were chosen such that the expected number of triggers per digitization was equal to the value in Eq. 3.10. This was done by solving

$$\frac{5,000,000}{n} \int_{P_{cutoff}}^{\infty} \frac{dP}{dp_n} dp_n = 4.59,$$

for the cutoff value, P_{cutoff}. For a frequency resolution of 1,000 mHz, a Gaussian distribution with a mean of 50 and standard deviation of $\sqrt{50}$ was used rather than an Erlang distribution for ease of computation. The power cutoffs were 16.04 σ, 29.64 σ, and 77.66 σ for resolutions of 40 mHz, 200 mHz, and 1,000 mHz, respectively. The cutoff for 1,000 mHz was changed to 82 σ during the analysis process. The original cutoff calculated using a Gaussian distribution underestimated the power, so more triggers than could be processed was produced.

A power spectrum was computed at each resolution for all 132,992 scans and the frequency and power of points above the respective thresholds were recorded for each resolution. This constituted the initial trigger list. The candidate list was formed by removing all non-axion like signals from the triggers list.

3.3.2 Data Quality Cut

The first cut removed triggers from digitizations when the experimental conditions were poor. This corresponds to times when the quality factor of the cavity was either below 10,000 or above 120,000 and when the system noise temperature was above 5 K. The lower bound of the Q cut removed digitizations where the power of an axion signal would not be large enough to be above background. The upper bound of the Q cut is due to that being the maximum Q value the cavity could have achieved, so anything larger was nonphysical. These cuts removed triggers from 3,206 digitizations.

3.3.3 Synthetic Axions

To test the experiment's detection capabilities, ADMX injected synthetic axion signals into the cavity, without the observers knowledge. These blind injections were coded to produce signals with the characteristic Maxwell-Boltzmann line shape expected from the isothermal halo model. The system is called the Synthetic Axion Generator (SAG) and these synthetic signals are referred to as SAGs. SAGs were not always being injected into the cavity. The system was turned off for part of the time so there were digitizations without SAGs present. Data taking was split up into nibbles, each covering approximately 10 MHz. At the end of a nibble, the observers presented a list of possible SAGs to the controllers to check if all were found.

A log of SAG frequencies was kept, however the metadata for digitizations does not contain information to say whether a specific digitization had a SAG present. Thus, an algorithm to determines whether a SAG was present in a digitization was written. At the finest resolution of 20 mHz, SAGs were immediately obvious in the power spectra. The SAGs were made up of approximately 30 components, each with a FWHM of about 60 mHz, 50 Hz apart.

Figures 3.12 and 3.13 show examples of a SAG, looking at the full spectrum and a zoomed in view. SAGs can be easily detected in HiRes by exploiting the fact the components are 50 Hz apart. Taking the frequency modulus 50 resulted in the SAG collapsing to a single peak. This is shown in Fig. 3.14.

The algorithm written to determine SAGs utilized this key feature. The first step found all triggers in a digitization. Second, the frequency modulus 50 was computed.

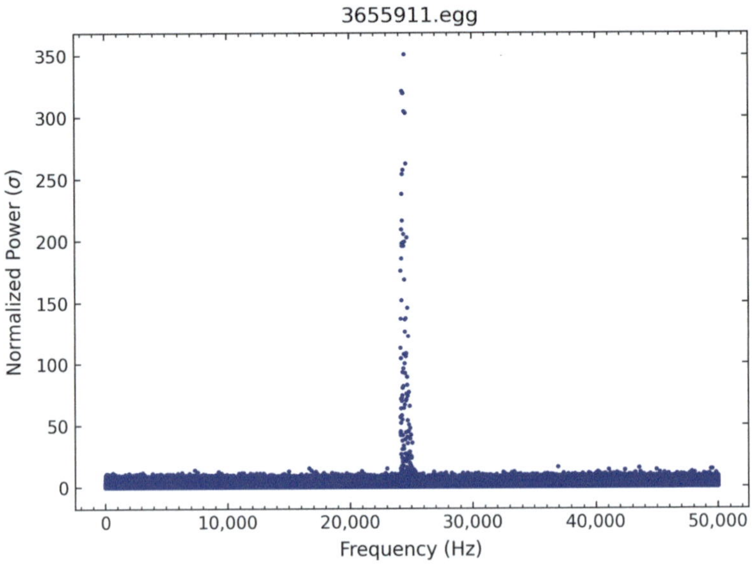

Fig. 3.12 Example SAG injection from ADMX run 1c

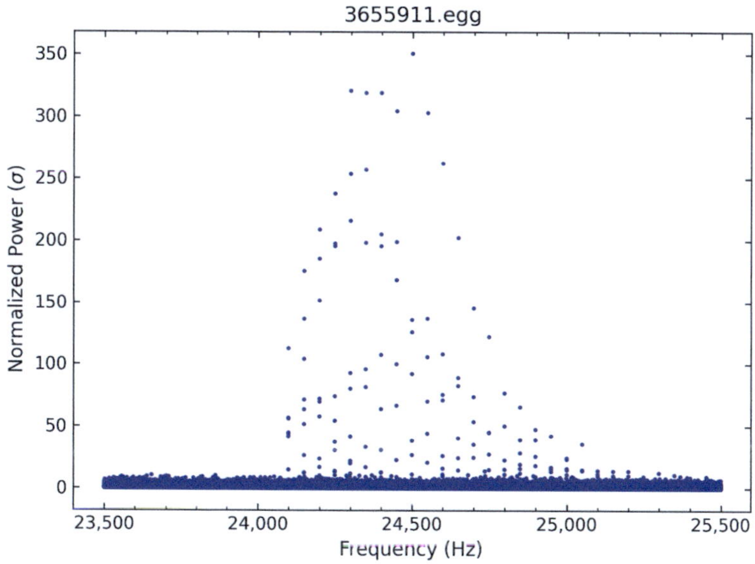

Fig. 3.13 Example SAG from run 1c, plotted around the peak

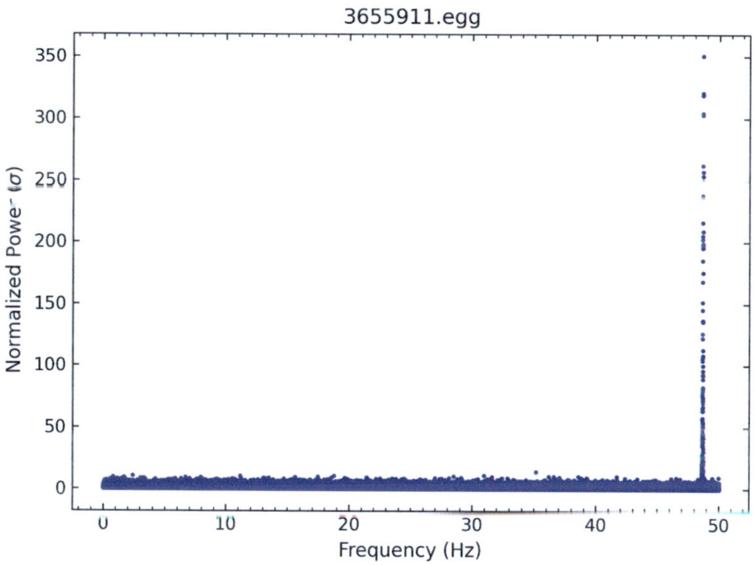

Fig. 3.14 Example SAG with frequency modulus 50, collapsing to a single peak

Next, the maximum power point was identified and all triggers within 150 mHz below to 150 mHz above are counted. The ratio of this count to the total number of triggers was taken. If the ratio was greater than 0.65, the digitization was flagged as containing a SAG.

This algorithm had a couple advantages; it could identify weak SAGs where only five of the thirty components were above background. It also allowed for only the SAG signal to be removed while saving other triggers in the digitization. The primary disadvantage was the algorithm could incorrectly identify a narrow signal. A narrow signal would contain fewer triggers than a SAG, so all digitizations flagged as a SAG with relatively few triggers was cross checked to confirm the result. In hindsight, requiring the triggers to span more 50 Hz would have yielded a proper result.

The first step in determining SAG files was to compile all digitizations containing a SAG frequency. Next, the above algorithm was run on this set. It identified 2,624 digitizations as containing a SAG. During the analysis process, it was discovered that several went unidentified for a few reasons. One cause was the width of the signal with frequency modulus 50 was spread across approximately 600 mHz rather than the 300 mHz expected. Another cause was weak SAGs containing about 10 triggers along with greater than five statistical triggers resulting in a failure. The others failed due to the peak frequency modulus 50 being wrapped around 0 and 50 Hz. Figure 3.15 depicts an example. The unidentified SAGs were added to the SAG list after their discovery. The cataloged SAG triggers were removed from the trigger list.

3.3.4 RFI

Another source of excess power in the experiment was radio frequency interference (RFI). This classification simply refers to non-axion like signals. The source of RFI was not always known as there are many devices which emit signals at the frequencies

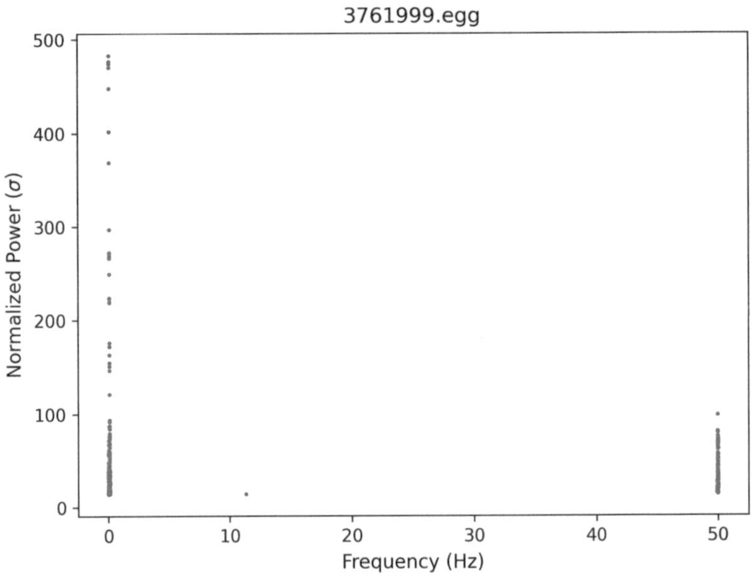

Fig. 3.15 An unidentified SAG, due to the frequency modulus 50 producing two peaks

probed. There were several tests to determine if a signal is RFI. Two in particular were enough to eliminate most RFI. One was an air check done at the ADMX site, during data taking, with an antenna to see if the signal could be detected outside of the cavity. Another was to check if the signal followed the Lorentzian of the cavity. A signal originating inside a cavity will decrease in power when the resonant frequency is tuned away from the signal frequency. If the signal power remains relatively constant that points to it having an external source.

A common source of RFI is electronics in the lab. There are effectively two categories; electronics which produce a constant signal and faulty electronics. During the early data taking, a significant amount of RFI was detected. The cause was found to be a defective power supply for one of the stepper motors and once replaced the RFI disappeared. These signals were easily identifiable and thus easily removable. Unfortunately, a significant number of digitizations were contaminated with this RFI. The contamination was present for an entire nibble, a frequency range of approximately 10 MHz. They contained two or three signals, spaced 20 kHz apart. Each of these were composed of many components, similar to SAGs, 60 Hz apart. Figures 3.16 and 3.17 depict an example.

The color coding shows the three components remain distinct once the frequency modulus 60 is computed. Although the signals were always spaced 20 kHz apart, their central frequencies would change from file to file. Unfortunately, these signals were not always observable in MedRes so a comparison could not be done. Given this contamination was spread across 10 MHz of data, the frequencies were not recorded in the ADMX ELOG, making it especially difficult to remove. The majority of

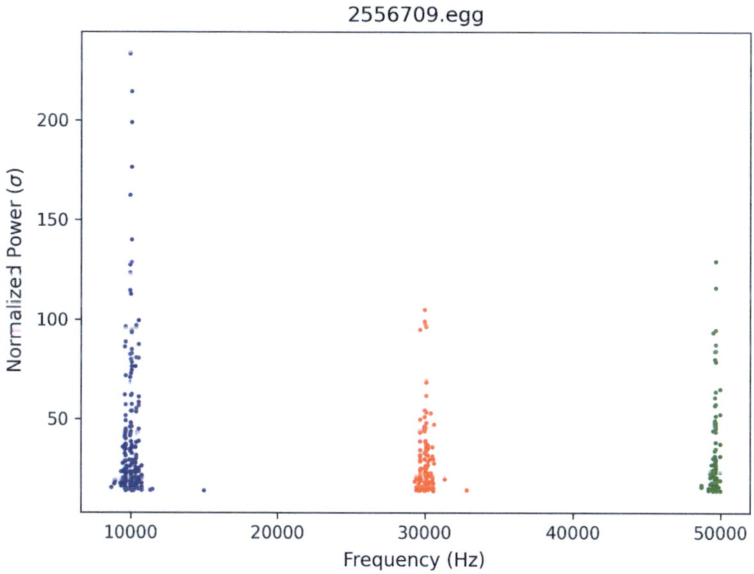

Fig. 3.16 Example of RFI in run 1c from a faulty power supply

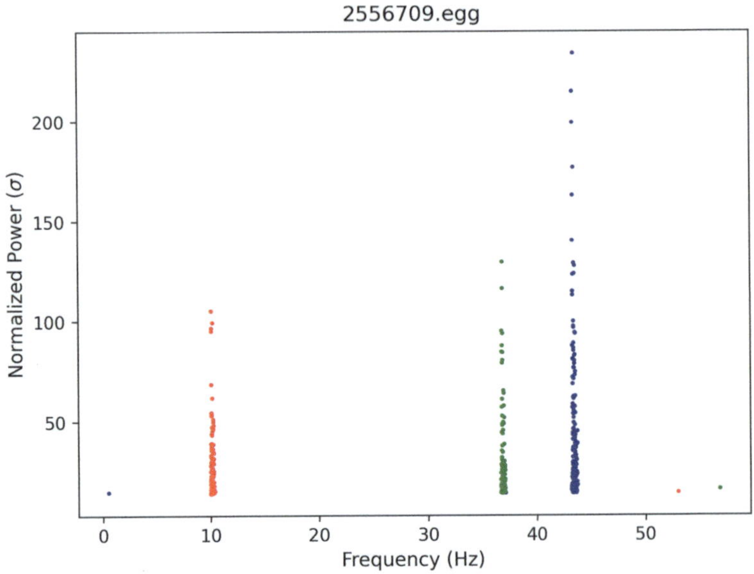

Fig. 3.17 RFI with frequency modulus 60

the digitizations containing these signals were found by examining files with many triggers, for example more than 100. Any missed digitizations were found later in the analysis process.

An axion signal from a narrow flow should produce only a few triggers in a single digitization, so other RFI signals were found by examining files with an excess number of triggers, between 20 and 100. Often times these signals were present in MedRes, so their frequencies were recorded in the ELOG, allowing them to be easily removed.

During the analysis process, digitizations containing RFI were logged. All triggers from these digitizations were removed. In total, 8,249 digitizations contained RFI visible to HiRes. Some digitizations contained RFI visible to MedRes, however at HiRes resolutions did not produce any triggers.

3.3.5 *Frequency Modulation*

One assumption made was the velocity vector in the rest frame of each cold, nonvirialized flow remains constant on the timescale of the experiment. Equation 3.1 gives the energy of an axion, where the velocity vector is with respect to the detector's velocity vector. As the Earth rotates around its axis and orbits the sun, this velocity is constantly changing. Thus, the frequency of photons produced by axions also changes with time. Figure 3.18 depicts the motion of the Earth within the galaxy.

Fig. 3.18 Diagram of Earth's orbit in the galaxy. Credit: http://www.sciencеminusdetails.com/2012/06/is-solar-system-tilted-sideways.html

The Earth's motion results in annual and diurnal frequency modulations of an axion signal. The energy of a photon produced from an axion can be written as [12]

$$hf = m_a c^2 + \frac{1}{2} m_a \vec{v} \cdot \vec{v}, \tag{3.11}$$

where h is Planck's constant, m_a is the axion's mass, and $\vec{v}$ is the velocity of the axion with respect to the detector. This velocity can be written as $\vec{v} = \vec{v}_a - \vec{v}_D$, where $\vec{v}_a$ is the velocity of the axion flow and $\vec{v}_D$ is velocity of the detector. The detector velocity can further be written as $\vec{v}_D = \vec{v}_\odot + \vec{v}_e + \vec{v}_U$, where $\vec{v}_\odot$ is the velocity of the sun around the galaxy, $\vec{v}_e$ is the velocity of the Earth as it orbits the sun, and $\vec{v}_U$ is the orbit of the detector at the University of Washington as the Earth rotates around its axis. The common frame amongst these is the galactic frame, so all velocities are given with respect to this. Equation 3.11 can be rewritten in terms of these new velocities

$$hf = m_a c^2 [1 + \frac{1}{2c^2} (\vec{v}_a - \vec{v}_\odot - \vec{v}_e - \vec{v}_U) \cdot (\vec{v}_a - \vec{v}_\odot - \vec{v}_e - \vec{v}_U)]. \tag{3.12}$$

Assuming the velocity of the sun around the galaxy is constant, the constants in Eq. 3.12 are m_a, $\vec{v}_a$, and $\vec{v}_\odot$. Terms containing only constants can be ignored, as they do not contribute to the frequency modulation. Also ignore terms only containing $\vec{v}_e$ and $\vec{v}_U$. This yields

$$hf = m_a c^2 \left[\frac{\vec{v}_\odot - \vec{v}_a}{c^2} \cdot (\vec{v}_e + \vec{v}_U) \right]. \tag{3.13}$$

Taking the derivative with respect to time gives us

$$\frac{df}{dt} = f_0 \left[\frac{(\vec{v}_\odot - \vec{v}_a)}{c^2} \cdot (\frac{d\vec{v}_e}{dt} + \frac{d\vec{v}_U}{dt}) \right], \tag{3.14}$$

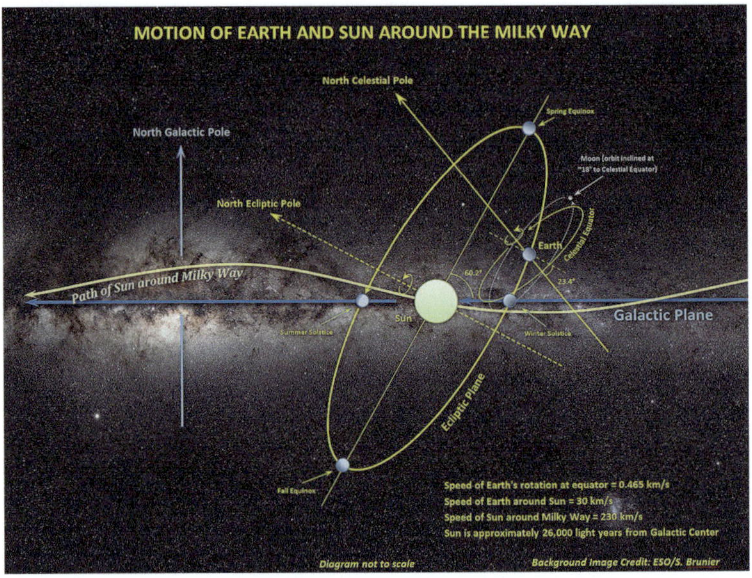

Fig. 3.19 Diagram of Earth's motion, in galactic coordinates.Credit: Jim slater307/Wikimedia Commons; background: ESO/S. Brunier

where $f_0 = \frac{m_a c^2}{h}$.

In order to calculate an approximate frequency modulation, the velocities at different times must be computed. Figure 3.19 provides a detailed view of Earth's motion in the galaxy with respect to the galactic frame.

For Earth's motion around the sun, a circular orbit was assumed. Working in cylindrical coordinates, $\hat{r}$ points along the galactic plane out from the center, $\hat{\phi}$ is also in the galactic plane perpendicular to $\hat{r}$, and $\hat{z}$ is normal to the galactic plane. The orbital plane of the Earth has an inclination of approximately 60° with respect to the galactic plane and is primarily in the $\phi - z$ plane. Labeling the magnitude of the Earth's velocity by v_e, the velocity of the Earth at any point in its orbit is given by

$$\vec{v}_e(\theta) = v_e \cos 60 \cos \theta \hat{r} + v_e \sin \theta \hat{\phi} + v_e \sin 60 \cos \theta \hat{z}, \tag{3.15}$$

where θ gives the position of Earth in its orbit, with $\theta = 0°$ when Earth is at the point labeled winter solstice and $\theta = 90°$ is when Earth is at the spring equinox point in Fig. 3.19.

For the detector, the axis of the Earth's rotation has an approximate inclination of 37° with respect to the galactic plane. Assuming the Earth's orbital plane and rotational plane are parallel

$$\vec{v}_U(\theta') = v_U \cos 37 \cos \theta' \hat{r} + v_U \sin \theta' \hat{\phi} + v_U \sin 37 \cos \theta' \hat{z}. \tag{3.16}$$

Approximate $\frac{d\vec{v}_e}{dt}$ and $\frac{d\vec{v}_U}{dt}$ with $\frac{\Delta\vec{v}_e}{\Delta t_e}$ and $\frac{\Delta\vec{v}_U}{\Delta t_U}$ over some chosen period of time. Using Eqs. 3.15 & 3.16

$$\frac{\Delta f}{\Delta t} = f_0 \left[\frac{(\vec{v}_\odot - \vec{v}_a)}{c^2} \cdot \left(\frac{\Delta\vec{v}_e}{\Delta t_e} + \frac{\Delta\vec{v}_U}{\Delta t_U} \right) \right]. \tag{3.17}$$

The sun is traveling in the $\hat{\phi}$ direction with a speed of 220 km/s. Earth has an orbital speed of 30 km/s. The detector is located in Seattle, Washington, with a latitude of 47.6062°, which yields a rotational speed of 0.31 km/s. Lastly, for the axion there are several choices. Table 3.1 is a list of example flow velocities in cylindrical coordinates [9].

The big flow, as mentioned before, is a well studied flow in the caustic ring model, infall with angular momentum flow is a generic flow, and the naive infall flow corresponds to axions falling in or out at approximately the escape velocity of the galaxy at the position of Earth.

To be conservative, the largest value of $\frac{df}{dt}$, which corresponds to the steepest slope of Eq. 3.14, was used. This was estimated using Eq. 3.17 for two points just before and after the steepest point.

First, the frequency modulation due to the Earth's orbit for a frequency of 1 GHz was calculated, which was approximately the largest frequency scanned in run 1c. Table 3.2 gives the maximum frequency modulation for each flow.

Similar calculations were done to estimate the frequency modulation from the Earth's rotation. When the Earth is at $\theta = 0$ in its orbit, the orbital plane and rotational planes are parallel, so the angles yielding a maximum shift are identical as before. Performing the same calculation yielded the frequency modulations given in Table 3.3.

Table 3.1 The velocities of example axion flows in cylindrical galactic coordinates

Flow	Velocity (km/s) $(\hat{r}, \hat{\phi}, \hat{z})$
Big flow	(120,505,0)
Infall with angular momentum	(0,140,635)
Naive infall	(618,0,0)

Table 3.2 Frequency modulations for each flow due to the Earth's orbit

Flow	Maximum frequency modulation (Hz/s)
Big flow	1.93×10^{-5}
Infall with angular momentum	3.69×10^{-5}
Naive infall	2.52×10^{-5}

Table 3.3 Frequency modulations for each flow due to the Earth's rotation

Flow	Maximum frequency modulation (Hz/s)
Big flow	7.42×10^{-5}
Infall with angular momentum	9.80×10^{-6}
Naive infall	1.35×10^{-4}

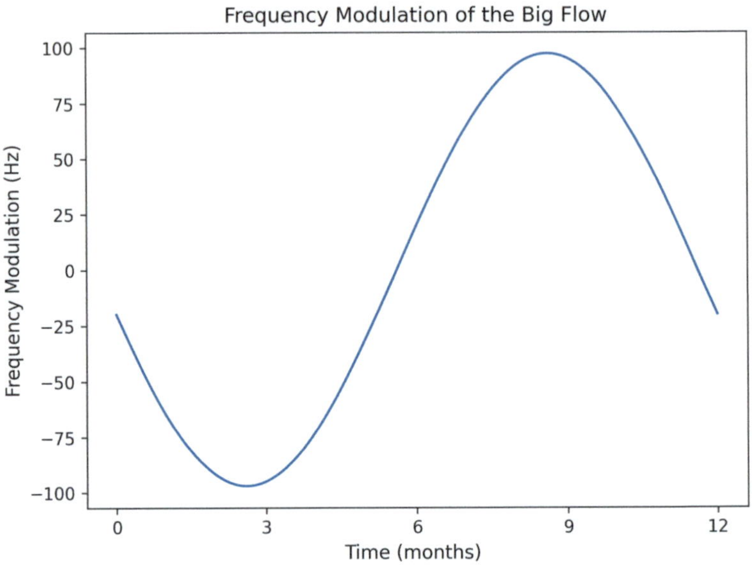

Fig. 3.20 Frequency modulation of the big flow due to Earth's orbit over one year

To calculate a frequency shift, these rates were multiplied by the elapsed time. If the time elapsed was greater than 12 h, only the frequency modulation due to the Earth's orbit was used. For less than 12 h time frames, the contributions due to Earth's orbit and rotation were summed.

Although Fig. 3.20 shows the largest frequency modulation is 200 Hz, recall ADMX broke up the frequency range into nibbles which only last approximately two weeks. Therefore, the largest frequency difference is about 3.69Hz/s $\times$ 14 days $\times$ 86400 s/day $\approx$ 48 Hz.

In the analysis, an excessively conservative value of a maximum frequency shift 100 Hz was chosen for the same time period. This frequency shift was used by partitioning the remaining triggers into clusters, where a cluster is a set of triggers with all adjacent triggers being 100 Hz of each another.

These clusters were used to perform two different cuts. First, clusters containing only a single trigger were immediately eliminated, referred to as a singleton cut. Second, for each trigger in a cluster, the expected frequency shift from the difference between timestamps was computed. If a trigger had no neighbors adjusted for time, it was eliminated. This left only clusters with an allowable frequency modulation. The next cut utilizing these clusters was a persistence cut.

3.3.6 Persistence

If an axion exists with mass m_a, Eq. 3.1 can be used to find the frequency of the resulting photon in the detector. Given enough power, the signal should be present each time the frequency is within the bandwidth of the cavity and the axion signal will persistent. The persistence of a cluster was calculated using

$$\gamma = \frac{k}{n}, \tag{3.18}$$

where γ is the persistence, n is the number of times the frequency of the cluster was within the bandwidth of the cavity, and k is the number of unique scans composing the cluster.

As discussed in Sect. 3.2.3, the background had fluctuating power, following an Erlang distribution. The power measured was signal+background, so it was possible to have an axion with power below the threshold be a trigger, if the background power was sufficiently large. In the following scan, if the background power fluctuated down and the frequency bin is no longer present, the signal appeared to be non-persistent. Due to this, a persistence cut of $\gamma = 1$ was too strict. A persistence cut of $\gamma = 0.3$ was used.

A persistence cut is dependable for frequency bins scanned many times. A bin width 100 Hz was sufficient in determining the number of times a frequency bin was scanned in run 1c. Unfortunately, the average number of scans per frequency bin was approximately six. Bins scanned only a single time were eliminated with the singleton cut. Many signals survived a first persistence cut on the clusters. However, a significant portion passed with, for example, a persistence of 1/3, 1/2, or 2/6. These ratios were qualitatively unreliable so an additional persistence cut was performed.

The second persistence cut was referred to as a first-half, second-half cut. By dividing the time signal of digitizations composing the frequency range of a cluster, the number of independent scans is doubled. The analysis process was repeated on this new set of scans. Dividing the time signal in half also doubles the frequency bin width in the power spectra, so the scans were zero filled to their original size, so exact frequency bins could be compared. The trigger threshold for each resolution was adjusted to account for the loss in SNR from a shorter integration time.

For resolutions of 20 mHz, 40 mHz, and 1,000 mHz, the first-half, second-half cut removed the remaining signals. For example, signals with $\gamma = 2/6$ now had $\gamma = 2/12$, so they could be eliminated. The 200 mHz resolution analysis still had surviving signals. They do not appear to be axion signals, further investigation needs to be completed. In particular, although these signals have allowable frequency modulations, they seem to not follow a smooth change. However, this must be confirmed by calculating the exact the motion of the Earth, and thus the frequency modulation, at the corresponding time. This work is ongoing by an ADMX collaborator.

3.4 Limit Setting

Two different limits were set. One on the axion-photon-photon coupling and another on the density of nonvirialized axion flows, assuming a DFSZ and KSVZ axion. The procedure for setting limits is simple. First, the threshold power selected for each frequency resolution and the persistence were used to determine an adequate effective threshold power. The effective threshold power will be in units of normalized power, σ, which must be converted to units of watts. Next, this value was set equal to Equation 2.22, which gives the expected axion power for certain experimental conditions.

Various losses to the axion power were considered. Finally, the desired limit was computed using a variance weighted mean, since each frequency bin has contributions from multiple digitizations. This ensured digitizations with experimental parameters far from the mean contribute less.

3.4.1 Power Threshold

An effective power threshold can be computed for a frequency bin using the threshold power, persistence, and combinatorics. Consider an axion signal at some frequency. Each time this frequency is scanned, the measured power is the sum of the axion power and the power from the fluctuating background. Scanning this frequency bin many times, one may expect the measured power to exceed threshold only in some subset of the scans, when the background power is high. The persistence cut demands a frequency bin has a trigger present present in some percent of the total number of scans. Recall the persistence cut used was $\gamma = 0.3$.

Ideally, limits are set using a power value as close to the axion power as possible, which will be called the effective threshold power. This was computed by subtracting a portion of power from the threshold, given by the expression

$$P_E = P_T - P_{bkg}, \tag{3.19}$$

where P_E, P_T and P_{bkg} are the effective threshold power, threshold power, and background power, respectively. Recall the thresholds were 13.9σ, 16.04σ, 29.64σ, and 82σ for frequency resolutions of 20 mHz, 40 mHz, 200 mHz, and 1,000 mHz, respectively. Next, how P_{bkg} is determined will be discussed.

The null hypothesis no signal survives the data cuts and thus no axion is detected. A signal would need to pass the persistence cut to invalidate the null hypothesis. The probability of this corresponds to a binomial distribution, where the probability of success is the probability a frequency bin is above threshold, which is labeled Λ. For a resolution of 20 mHz, the probability of a bin having a power greater than Λ is given by

$$p = \int_{\Lambda}^{\infty} e^{-x} dx = e^{-\Lambda}. \tag{3.20}$$

If a frequency bin is scanned n times, then a trigger must be detected in at least $m = \lceil \gamma n \rceil$ scans, where γ is the persistence cut value and the product is in the ceiling function. The total probability a signal will pass persistence is the sum of all possible successful outcomes, given by

$$\sum_{i=m}^{n} \binom{n}{i} p^i (1-p)^{n-i}. \tag{3.21}$$

To validate the null hypothesis at a 95% CL, the value of Λ allowing the background to produce a persistent signal with a probability of 95% must be found. To do so, Eq. 3.21 was set equal to 0.95, solved for p, then Eq. 3.20 was used to determine Λ. This was the value used for P_{bkg} in Eq. 3.19. If an axion signal had a power higher than P_E, the signal would have had a 95% chance of passing persistence, thus invalidating the null hypothesis. This can be more clearly seen with an example.

Suppose a frequency bin is scanned 20 times. how does the effective threshold change as the persistence cut becomes stricter? Table 3.4 shows the calculated p, P_{bkg}, and P_E values for a set of m values.

The stricter persistence cuts result in higher effective thresholds. The less strict the persistence cut, the more likely it is a for a signal to pass it. Thus, the axion power required to pass persistence goes down. However, the caveat is a more lenient persistence cut may not remove all signals, so the null hypothesis would be invalidated.

Intuitively, better limits should be achieved the more times a frequency bin has been scanned. Table 3.5 shows the result repeating the above example for a bin scanned 50 times.

Comparing the values in Tables 3.4 and 3.5, increasing the number of scans improves the limits. As the number of scans is increased, there are more combinations that pass persistence. Thus, the probability of a single event passing must go down, which is equivalent to raising Λ.

The above process assumed a single frequency bin, however an axion signal's frequency will modulate due to the Earth's motion. The combinatorics change slightly

Table 3.4 Effect of persistence cut on background contribution to power

m	p	P_{bkg}	P_E
2	0.216	1.53	12.37
6	0.456	0.79	13.11
8	0.558	0.58	13.32
12	0.741	0.30	13.60
16	0.896	0.11	13.79
20	0.997	0.003	13.90

Table 3.5 Contribution of background for a bin scanned many times

m	p	P_{bkg}	P_E
5	0.174	1.75	12.15
15	0.403	0.91	12.99
20	0.506	0.68	13.22
30	0.699	0.36	13.54
40	0.871	0.14	13.76
50	0.999	0.00	13.90

if a trigger is allowed to be within some number of bins of the original, depending on how much time has passed. This was remedied by treating each cluster of triggers as a single bin, rather than the individual bins composing it.

To correctly apply this, a proper count of how many times each frequency was scanned was needed. Dividing the frequency range 100 Hz bins resulted in correct counts. The number of times a frequency bin was counted corresponds to how many digitizations it was in the bandwidth of the cavity. Although the Fourier frequency for a digitization spans 50 kHz, the bandwidth could span from as little as 10 kHz centered on the resonant frequency to larger than the 50 kHz range.

Using the number of counts, $\gamma = 0.3$, and the thresholds for each resolution, the effective threshold power was computed for 100 Hz bin across the entire frequency range. For the resolutions coarser than 20 mHz, Eq. 3.20 was replaced with its corresponding PDF.

3.4.2 Power Conversion

Computing the effective threshold power yielded a value in units of normalized power, σ. To equate this to an expected axion power, it was put into units of watts. The Dicke radiometer equation provided the correct conversion [13]

$$P_N = k_b T_{sys} b_w, \tag{3.22}$$

where P_N is the rms noise power, which was simply the mean of the power spectrum for each resolution, k_b is Boltzmann's constant, T_{sys} is the system noise temperature, and b_w is the the frequency resolution. The rms noise power and b_w are both proportional to the number of sub-spectra, n, so the conversion from 1σ of normalized power to watts was the same for all resolutions. The standard deviation of power scales as $\sqrt{n}$, which effected the limits for varying resolutions.

3.4.3 Power Losses

Recall the expected axion power for the ADMX run 1c parameters, given by Eq. 2.22. An experiment does not detect 100% of this power, as there are losses from three main sources. First, the pick-up antenna only detects some fraction of the photons put into the cavity by the axions. This fraction is given by

$$\nu = \frac{\beta}{1+\beta},$$

where β is the coupling constant. At critical coupling $\beta = 1$, so the fraction detected is 1/2. However, the antenna is not always critically coupled. The coupling constant is measured for each digitizations, so this fraction can be computed for each.

Next, as the cavity is tuned through a signal's frequency, the power of the signal will follow a Lorentzian, with the maximum power being when the resonant frequency is at the signal frequency. For a given digitization, if the signal is slightly off resonance the detected power will be less than the full signal power. The Lorentzian of a resonant cavity is given by

$$L(f) = \frac{1}{1 + 4Q_L^2(1 - \frac{f}{f_0})^2}, \tag{3.23}$$

where Q_L is the loaded quality factor. To account for this loss, an effective Lorentzian factor, $\bar{L}$, was used [7]. For scans where the FWHM is smaller than the cavity bandwidth, the Lorentzian over the entire FWHM was given by

$$\bar{L} = \frac{Q_L}{f_0}\int_{f_0(1-\frac{1}{2Q_L})}^{f_0(1+\frac{1}{2Q})} L(f)df = \frac{\pi}{4}.$$

For digitizations where the quality factor was small, the FWHM may be larger than the cavity bandwidth, so only the cavity bandwidth was integrated over, which was approximately 24 kHz

$$\bar{L} = \frac{Q}{f_0}\int_{f_0-12\text{kHz}}^{f_0+12\text{kHz}} L(f)df = \frac{f_0}{24\text{ kHz}Q_L}\arctan\frac{24\text{ kHz}Q}{f_0}.$$

The FWHM of the cavity was recorded for every digitization, so this loss factor was computed for each.

The final source of power loss is a property of the discrete FFT algorithm [7]. The power spectra are composed of a finite number of frequency bins. Even if a signal is contained within a single bin, its power will be divided amongst several bins. However, the signal frequency may not be centered in a single bin, it could be near the edge of a bin. For a signal at f_s, with a frequency offset of Δ from a basis frequency f_b, the power in the f_b frequency bin is given by

$$P_b = \left(\frac{\sin\pi\Delta}{\pi\Delta}\right)^2.$$

The mean power in the central bin is calculated by integrating Δ over the bin width

$$P_b = \int_{-1/2}^{1/2}\left(\frac{\sin\pi\Delta}{\pi\Delta}\right)^2 d\Delta \approx 0.7737.$$

The bin width drops out, which is expected as this loss is a property of the discrete FFT independent of resolution.

The overall loss factor is given by

$$\varepsilon = \bar{P}_b \bar{L} \nu = \frac{0.7737 \bar{L} \beta}{1 + \beta}, \tag{3.24}$$

where the mean Lorentzian factor depends on the FWHM and cavity bandwidth. The final equality used to calculate the limits is given by

$$P_E P_N = \varepsilon P_a, \tag{3.25}$$

where P_E is the expected threshold power, P_N is given by Eq. 3.22, and P_a is given by Eq. 2.22. Equation 3.25, along with the meta data from each digitization, was used to compute the desired limits.

3.4.4 Weighted Averaging

The run 1c frequency range was divided 100 Hz bins and the limits for each were computed by averaging the results from the contributing digitizations. Each digitization had varying meta data values, such as the loaded quality factor. This resulted in the standard deviation of power being different scan to scan. This was accounted for by performing a variance weighted mean.

The variance for a single scan was labeled by σ_i^2, so the weights are given by $w_i = \frac{1}{\sigma_i^2}$. The weighted mean is then computed using

$$\bar{X} = \frac{\sum_{i=0}^{i=N} w_i x_i}{\sum_{i=0}^{i=N} w_i},$$

where N is the number of digitizations contributing to frequency bin. The limit value for a single scan, x_i, is computed using

$$x_i = \frac{P_{E,i} P_{N,i}}{\varepsilon P_{a,i}}. \tag{3.26}$$

The standard deviation for a scan is computed using

$$\sigma_i = \frac{\sqrt{n} P_{N,i}}{\varepsilon P_{a,i}}, \tag{3.27}$$

where n is the number of sub-spectra, corresponding to the desired frequency resolution. For the axion density limit, Solving Eq. 3.25 for ρ_a, yields

$$\begin{aligned} x_i &= \rho_{a,i} \\ &= \frac{P_{E,i} P_{N,i}}{\varepsilon P_{a,i}} \\ &= \frac{P_{E,i} \times kT_{sys,i} b \times (1+\beta)}{0.7737 \bar{L} \beta \times P_{a,i}}, \end{aligned} \tag{3.28}$$

where the axion power is given by

$$P_{a,i} = 7.7 \times 10^{-23}\,\mathrm{W} \left(\frac{\mathrm{V}_i}{136\,\ell}\right) \left(\frac{\mathrm{B}_i}{7.5\,\mathrm{T}}\right)^2 \left(\frac{\mathrm{C}_i}{0.4}\right) \left(\frac{g_\gamma}{0.36}\right)^2 \left(\frac{\mathrm{f}_i}{1\,\mathrm{GHz}}\right) \left(\frac{\mathrm{Q_{L,i}}}{80,000}\right), \tag{3.29}$$

where $g_\gamma = -0.97(0.36)$ for KSVZ (DFSZ). The standard deviation is given by

$$\sigma_i = \frac{\sqrt{n} \times kT_{sys,i} b \times (1+\beta)}{0.7737 \bar{L} \beta \times P_{a,i}}. \tag{3.30}$$

To set limits on the axion-photon-photon coupling, and its degeneracy with the density, Eq. A.5 was solved for g_γ and substituted into the Eq. 3.29. For the coupling alone,

$$\begin{aligned} x_i &= g_{a\gamma\gamma,i} \\ &= \sqrt{\frac{P_{E,i} \times kT_{sys,i} b \times (1+\beta)}{0.7737 \bar{L} \beta \times P_{a,i}}}, \end{aligned} \tag{3.31}$$

where now the axion power is given by

$$P_{a,i} = 5.68 \times 10^{26}\,\mathrm{W} \left(\frac{\mathrm{V}_i}{136\,\ell}\right) \left(\frac{\mathrm{B}_i}{7.5\,\mathrm{T}}\right)^2 \left(\frac{\mathrm{C}_i}{0.4}\right) \left(\frac{\rho_{a,i}}{f \times 1\,\mathrm{GHz}}\right) \left(\frac{\mathrm{Q_{L,i}}}{80,000}\right), \tag{3.32}$$

and the density can be set to any desired value. The standard deviation given by Eq. 3.30 and the axion power by Eq. 3.32 were substituted into Eq. 3.31.

3.5 Run 1c HiRes Limits

The limits on density were computed using the axion-photon-photon coupling for both KSVZ or DFSZ. Figure 3.21 shows the limits for each frequency resolution.

The hairline gaps are TE mode crossings, which prevented quality data from being taken. Although the limits for finer resolutions seem better across the entire range, this is only true if the frequency dispersion of the axion signal is approximately the bin width. The four main flows studied in the caustic ring model are the big, little, up, and down flows. The densities of the big, little, up, and down flows were estimated to be 11 GeV/cm^3, 1.1 GeV/cm^3, 5.3 GeV/cm^3, and 4.7 GeV/cm^3, respectively [9].

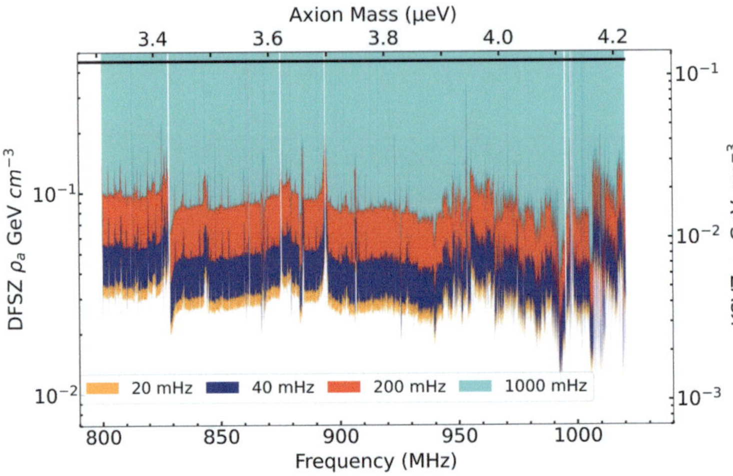

Fig. 3.21 The limits on axion flow density. The left axis assumes DFSZ coupling, while the right assumes KSVZ. The horizontal line is at 0.45 GeV for DFSZ coupling

Assuming the caustic ring model to be correct, these results confidently rule out the DFSZ axion, and thus any axion with a larger coupling. Assuming instead the existence of the DFSZ axion, a stringent limit is placed on the density of nonvirialized axion flows.

The next set of limits are on the more familiar axion-photon-photon coupling, shown in Fig. 3.22. A flow density of 0.45 GeV was used, as that is the average

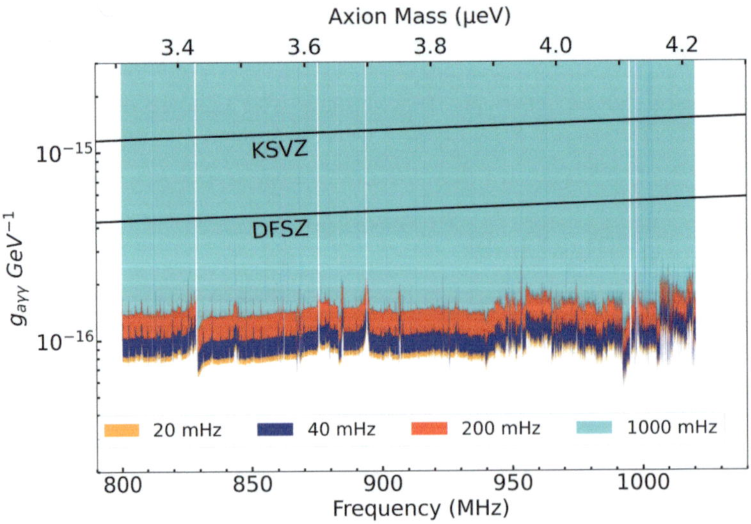

Fig. 3.22 The limits on axion-photon-photon coupling. The coupling predicted by the KSVZ and DFSZ models are shown

density of the dark matter halo. It is important to note this density value could vary by as much as two orders of magnitude for nonvirialized flows. Again, the limits rule out DFSZ axions across the entire frequency range. For the DFSZ axion, the density of a nonvirialized flow would need to be less than approximately 0.016 $\mathrm{GeV/cm^3}$ to have gone undetected. Similarly, for KSVZ sensitivity the density would need to be less than about 0.0021 $\mathrm{GeV/cm^3}$.

References

1. P. Sikivie, Phys. Lett. B **432**, 139 (1998). https://doi.org/10.1016/S0370-2693(98)00595-4
2. P. Sikivie, Phys. Rev. D **60**, 063501 (1999). https://doi.org/10.1103/PhysRevD.60.063501
3. L.D. Duffy, P. Sikivie, Phys. Rev. D **78**, 063508 (2008). https://doi.org/10.1103/PhysRevD.78.063508
4. L.D. Duffy et al., Phys. Rev. D **74**, 012006 (2006). https://doi.org/10.1103/PhysRevD.74.012006
5. J.D. Hoskins, A modulation sensitive search for non-virialized axions in the Milky Way halo, Ph.D. thesis, University of Florida 2014
6. J. Hoskins et al., Phys. Rev. D **94**, 082001 (2016). https://doi.org/10.1103/PhysRevD.94.082001
7. J. Hoskins et al., Phys. Rev. D **84**, 121302 (2011). https://doi.org/10.1103/PhysRevD.84.121302
8. C. Bartram et al., Non-virialized axion search sensitive to doppler effects in the milky way halo (2023). Preprint at arXiv:2311.07748 [astro-ph]
9. S.S. Chakrabarty, Y. Han, A.H. Gonzalez, P. Sikivie, Phys. Dark Universe **33**, 100838 (2021). https://doi.org/10.1016/j.dark.2021.100838
10. Y. Semertzidis, J.D. Vergados, (2015). Preprint at arxiv:1511.08516 (publisher: arXiv Version Number: 1)
11. F.S. Ling, P. Sikivie, S. Wick, Phys. Rev. D **70**, 123503 (2004). https://doi.org/10.1103/PhysRevD.70.123503
12. D.B. Tanner, Expected doppler shifts (2021) (unpublished)
13. R.H. Dicke, Rev. Sci. Instrum. **17**, 268 (1946). https://doi.org/10.1063/1.1770483

Chapter 4
ADMX Prototyping

Abstract One incredible hurdle in the search for axions is that the allowable mass range spans more than 12 orders of magnitude, where each band requires new resonators and detection schemes to be developed. This chapters discusses the development of a new cavity for ADMX to search for axion dark matter in a frequency range of 2 to 4 GHz. The primary feature optimized in this work is unloaded quality factor of prototype cavities. Additionally, design and testing of impedance filters is discussed. This work results in a cavity design with an unloaded quality factor of approximately 10,000, which matches simulations. These results are a promising first step in developing a detector with QCD axion detection capabilities.

4.1 Development of Future Experiments

One of the downsides to a cavity haloscope search is the relatively limited tuning range. ADMX has scanned approximately 650 MHz to 1.5 GHz using a single cavity with a volume of approximately 136 L, by adjusting the tuning rod set-up. Searching for higher mass axions means the frequency increases, requiring a new, smaller cavity. ADMX is developing a next-generation detector to scan 2 to 4 GHz.

This frequency range requires a cavity with an inner diameter of approximately 13 cm. Estimating a length of 1 m, this cavity would have a volume of approximately 13 L. If a single cavity was used, the expected axion power would decrease by a factor of 10, as compared to ADMX run 1c. Solving for the axion-photon photon coupling in Eq. 2.19, the volume dependence is $V^{1/2}$ and the time dependence is $t^{1/4}$, so a factor of 10 decrease in volume requires scanning for 100 times longer to hit the same sensitivity. Thus, a multi-cavity array which utilizes the entire volume of the magnet is required to achieve appreciable results.

ADMX EFR will be an 18-cavity array, a diagram of which is shown in Fig. 4.1. The experiment will be housed at Fermilab and and utilize a repurposed MRI magnet with a magnetic field strength of approximately 9.2 T. A rendering of the experiment is shown in Fig. 4.2. The 18-cavity array requires more electronics, so the experiment will utilize two cryogenic systems; one system for the cavity and magnet and a second

A. Hipp, *Searching High and Low for the Axion*,
Springer Theses, https://doi.org/10.1007/978-3-032-02276-9_4

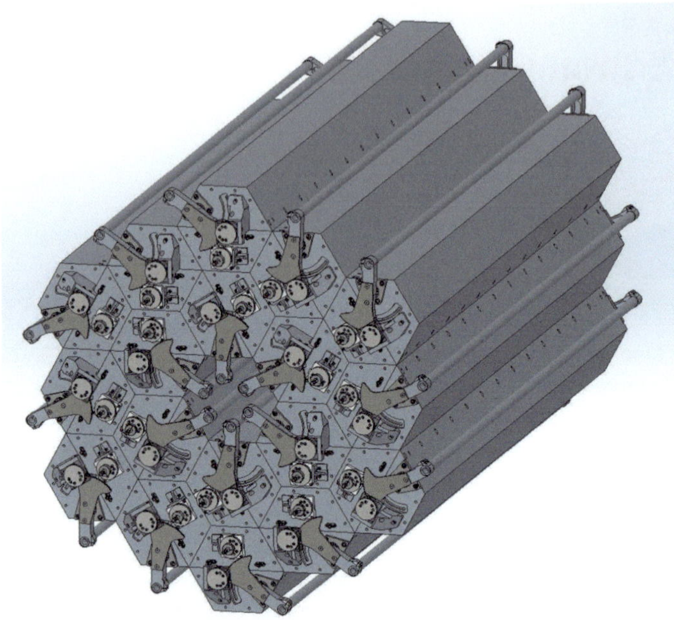

Fig. 4.1 Diagram of the 18-cavity array

Fig. 4.2 Rendering of the EFR experiment at FNAL

for the electronics. The signal will be transferred from the cavities to the electronics via low-loss coaxial line, which is still being developed.

Transitioning to a multi-cavity experiment presents new challenges. One is the cavities must be frequency locked, so their power spectra are centered on the same resonant frequency. Another challenge is determining how to properly add the power from each cavity. The research and development of ADMX EFR is divided amongst

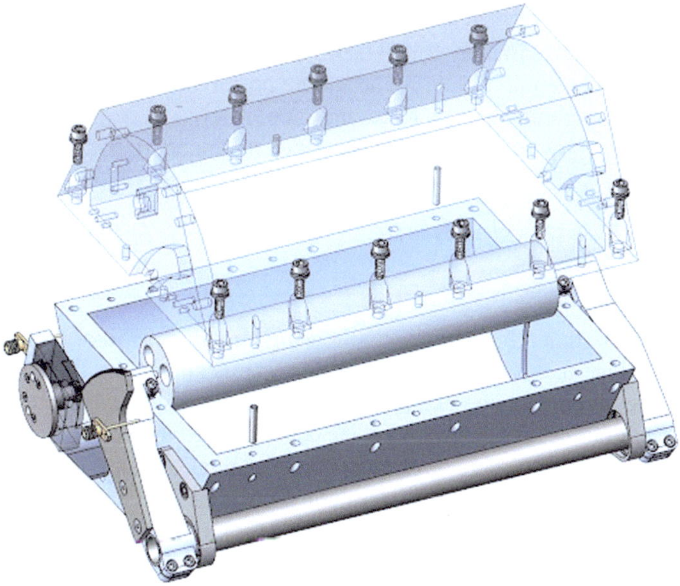

Fig. 4.3 Diagram of the version 3 cavity

several of the collaboration members. The work done at University of Florida focuses on the optimization of the quality factor for the EFR cavities, the results of which will be presented.

4.2 Cavity Design

The cavities have a clam-shell design, with an inner diameter of approximately 13 cm, as shown in Fig. 4.3. This design is easy to assemble and align properly, making the set up of the final 18-cavity array substantially easier. The cavities used for data taking will be made of copper and have a length of approximately 1 m, however the prototypes tested were made of Aluminum 6061 and had a length of approximately 20 cm. This reduced the cost of building prototypes and made tabletop measurements easier.

4.3 Quality Factor Optimization

Recall from Eq. 2.23, the scan rate for an axion search is proportional to the loaded quality factor, Q_L. Optimizing the quality factor will reduce the total time needed to scan the 2 GHz range. As frequency ranges between experiments begin to overlap, being able to scan quickly is important.

In total, four prototype cavities were built and tested. The prototypes had a frequency range of approximately 2 to 3 GHz. The designs were labeled by a version number, where only versions 3, 5, 7, and 8 were prototyped. All measurements were done at room temperature, using a Keysight E5063A VNA. A definition of the cavity quality factor and how to measure it is described in Appendix B.

4.4 Cavity Sweeps

The cavities were tuned both by hand and using an Attocube rotary motor. Tuning by hand was done in two ways. Each cavity was machined with three blind holes, such that a metal pin could be placed through the tuning mechanism into each hole. The three positions set the tuning rod at the center of the cavity, half-way tuned, and at the wall of the cavity. This allowed for quick initial measurements to be made of new prototypes. The version 3 and version 5 cavities were tuned using a dummy motor, which was turned by hand. This allowed for a larger number of points to be measured.

Additional measurements of version 5 and all following prototypes were tuned using an Attocube ANR240 rotary motor. A script was written using Python to automate the data taking process. The script sent instructions to the rotary motor to tune the cavity and to the Keysight VNA to record the S_{21}, S_{11}, and S_{22} spectra, saving the data to an hdf5 file. Five spectra spanning 1.9 to 3.1 GHz were captured to ensure the point spacing was fine enough for an accurate Q calculation.

The number of Q measurements taken corresponded to the number of tuning steps. This was set by a sweep parameter which specified how many tuning steps to take to sweep the entire 2 to 3 GHz frequency range. Two types of sweeps were performed; a coarse tuning with a larger frequency spacing and a fine tuning with relatively small frequency spacing. The coarse tuning took significantly less time to complete, so this was primarily done to compare the Q between different configurations of a particular version. For example, comparing the performance of PEEK tuning arms versus aluminum tuning arms. These comparisons only required a general trend of the Q across the tuning range to be found. The fine tuning was done to resolve mode crossings. Mode crossings occur when the TM modes pass through and interfere with TE modes. For some cavity configurations, mode crossings were especially sharp, such that the coarse tuning sweeps completely skipped over them. Resolving the TE modes is important as they are regions where science data cannot be collected.

4.4.1 *Cavity Version 3*

The version 3 cavity had the aforementioned dimensions and was tuned with a 3.5 cm diameter aluminum tuning rod. The quality factor versus frequency of the cavity was measured using a coarse frequency resolution, by manually tuning the cavity with

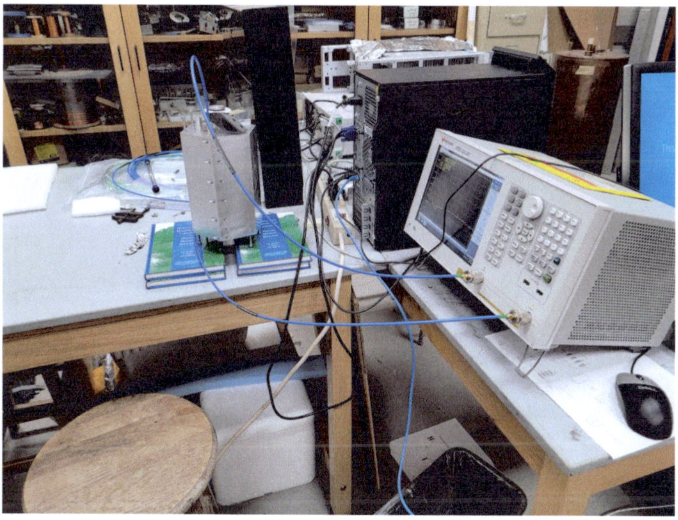

Fig. 4.4 Set-up of the cavity measurement process

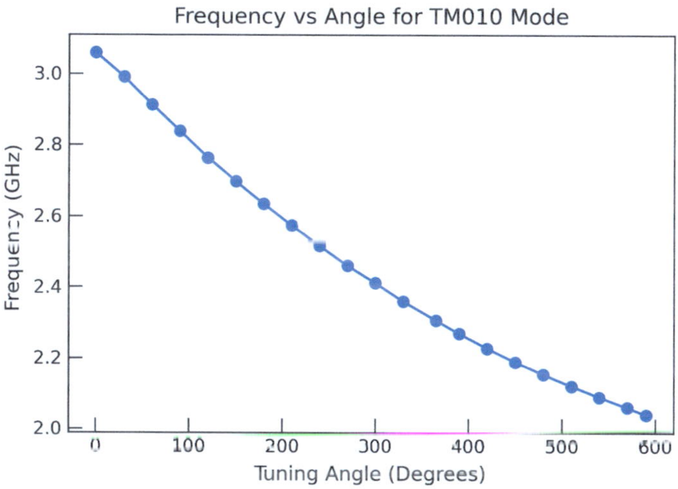

Fig. 4.5 Frequency versus tuning for the TM_{010} mode

a dummy motor. The set-up used to test the cavity is shown in Fig. 4.4, where two coaxial cables were attached to ports on opposite ends of the cavity then to the two ports on the VNA.

The first cavity property measured was the frequency versus tuning range, shown in Fig. 4.5, to confirm the predicted tuning range.

An example S21 spectrum is shown in Fig. 4.6, where the lowest frequency mode is the TM_{010} mode. Initially a 5 mil Kapton gasket was used to separate the two

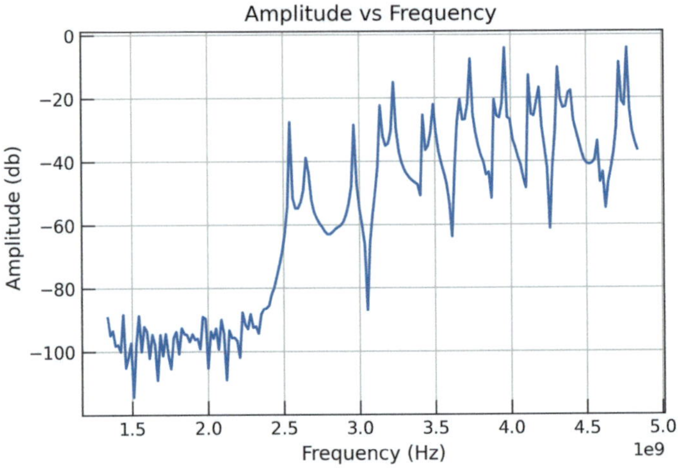

Fig. 4.6 S21 power spectrum for the version 3 cavity

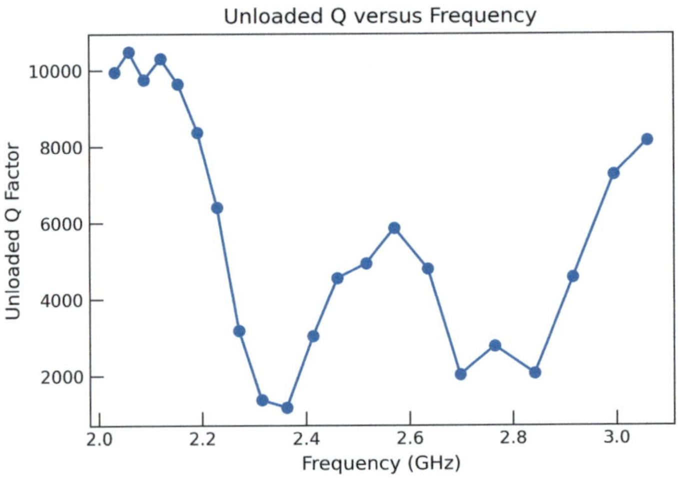

Fig. 4.7 Unloaded quality factor for the version 3 cavity with a 5 mil Kapton gasket

halves of the clam-shell. It was thought that separating the halves would spoil the TE modes, since they produce currents in the phi-direction of the cavity. TE modes interfere with the TM_{010} mode, resulting in frequency ranges where data cannot be collected. The Q versus frequency for this configuration is shown in Fig. 4.7.

Equation B.2 shows the quality factor is proportional to the frequency of the mode, so Q should fall off smoothly across the range. The presence of the gasket is visibly detrimental to the quality factor. Rather than spoiling the TE modes, it was hypothesized the Kapton gasket instead acted as a waveguide, introducing large interfering modes. Removing the gasket and remeasuring the quality factor yielded

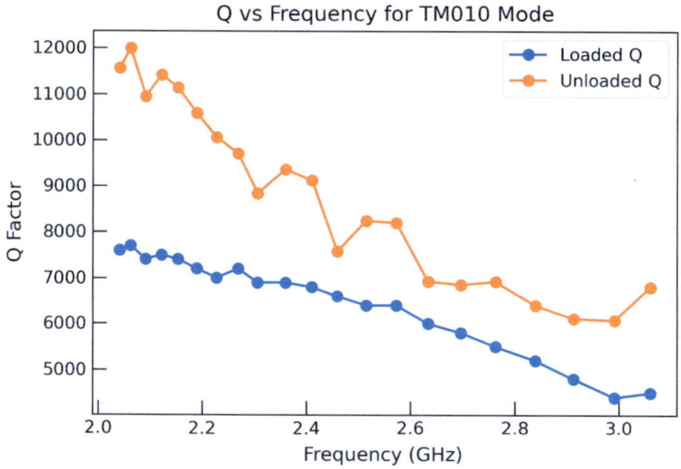

Fig. 4.8 Loaded versus unloaded Q for the version 3 cavity

the results shown in Fig. 4.8. The Kapton gasket configuration was scrapped based on these results.

The unloaded quality factor for the new configuration was lower across the entire frequency range as compared to simulations. It was found that the surface finish of the cavity interior was moderately rough. The average surface roughness and maximum surface roughness was measured. The worst average surface roughness was approximately 60 µin and the maximum was 464 µin. An ADMX collaborator, Mark Jones, simulated the cavity, varying the surface roughness, and computed the unloaded quality factor for the tuning rod on center. The results of the simulation are shown in Table 4.1.

The closest value measured was at a frequency of 3.060 GHz, which had $Q_U = 6,807$. This value matches up well between the simulation and the measured surface roughness. The subsequent prototypes had similar surface roughness profiles, so the effect to the quality factor was taken into account.

It was determined the version 3 cavity was slightly too large, such that the 18-cavity would not fit inside the magnet bore, so version 5 was designed with a smaller diameter.

Table 4.1 Effect of surface roughness on Q_U

Surface roughness (μ in)	Frequency (GHz)	Q_U
0	3.098	9,168
30	3.098	8,136
60	3.098	7,017
120	3.098	6,502

4.4.2 Cavity Version 5

The version 5 cavity had an inner diameter of 12.8 cm and used a tuning rod with a diameter of 3.1 cm. In addition, there was a rounded transition for the cylinder walls to end plates. A diagram of version 5 is shown in Fig. 4.9.

The quality versus frequency is shown in Fig. 4.10. The large dip just above 2 GHz is due to interference between the TM_{010} mode and a TE mode. Mode crossings are ignored at this stage, since the frequency of TE modes is dependent on the length of the cavity. Given the prototype was much shorter than the expected final length, the location of the TE modes will shift.

Unlike in version 3, where the quality factor decreased relatively smoothly over the tuning range, the quality factor starts high, drop significantly, then recovers at higher frequencies. It was hypothesized that the change in the ratio between the cavity inner diameter and tuning rod diameter resulted in the mode distribution changing, such that more power was leaking out of the slot.

To determine the effect of power leakage out of the slot, Q_U was measured for the empty cavity, with the tuning rod removed, then again after covering the slots with aluminum tape. The TM_{010} mode had a frequency of approximately 1.79 GHZ for the empty cavity. With open slots, $Q_U \simeq 11,900$ and with the slots covered with aluminum tape, $Q_U \simeq 20,300$. These measurements confirmed the slots resulted in increased energy loss.

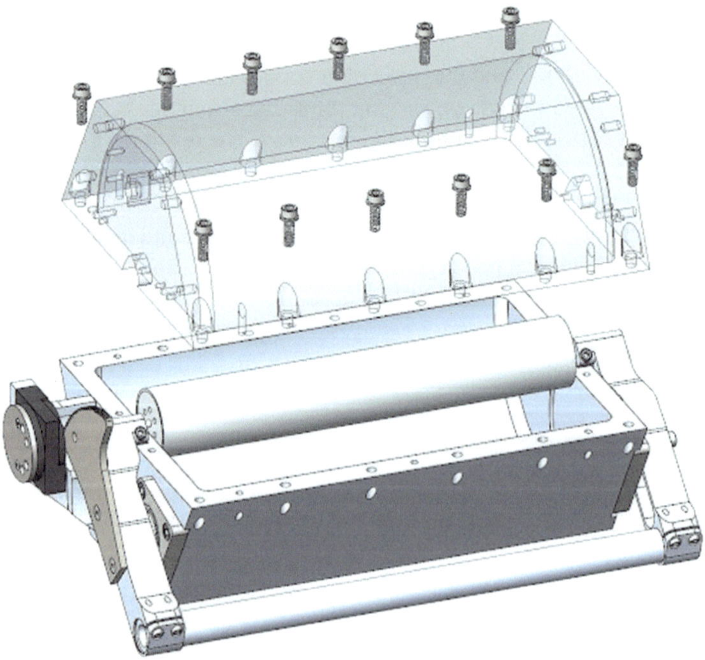

Fig. 4.9 Diagram of the version 5 cavity

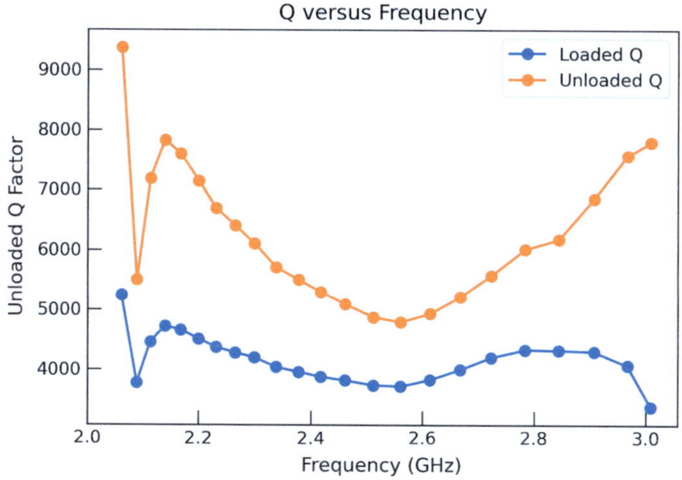

Fig. 4.10 Loaded versus unloaded Q for the version 5 cavity

The subsequent design, version 7, was changed to have an internal tuning mechanism, removing the slots.

4.4.3 Cavity Version 7

The version 7 cavity had the same dimensions and utilized the same tuning rod as version 5. A diagram of version 7 is shown in Fig. 4.11 and a diagram of its tuning mechanism is shown in Fig. 4.12.

Moving to an internal mechanism meant introducing additional material inside the cavity; an arm holding the tuning rod and an axle connecting the arm to the exterior of the tuning mechanism were present in the cavity. Arms, screws, and axles made of PEEK and aluminum were tested. The first measurements made were for the empty cavity with different configurations of materials, shown in Table 4.2.

The quality factor of the empty V7 cavity is 57% higher than that of version 5, so removing the slots and moving to an internal tuning mechanism did result in less energy loss. Although introducing the axles, screws, and arms lowers the quality factor, even the worst case of all PEEK parts performs better than version 5.

Several quality factor measurements were done, again using various configurations of the parts. The best performing configuration was using aluminum for all three parts. A plot showing the unloaded quality factor versus frequency for this configuration is shown in Fig. 4.13.

Comparing these results to those of version 5, although the quality factor increased across the entire range, it was still quite poor. Although removing the slot decreased the energy loss, it was thought the axle of the tuning rod was still coupling power

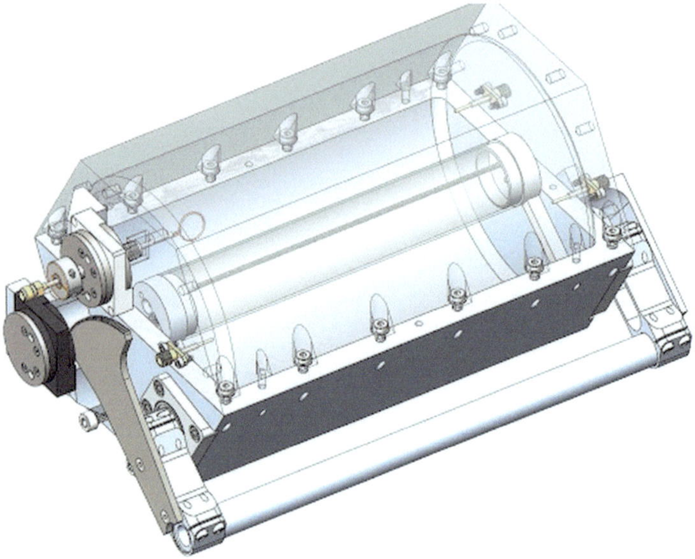

Fig. 4.11 Diagram of the version 7 cavity

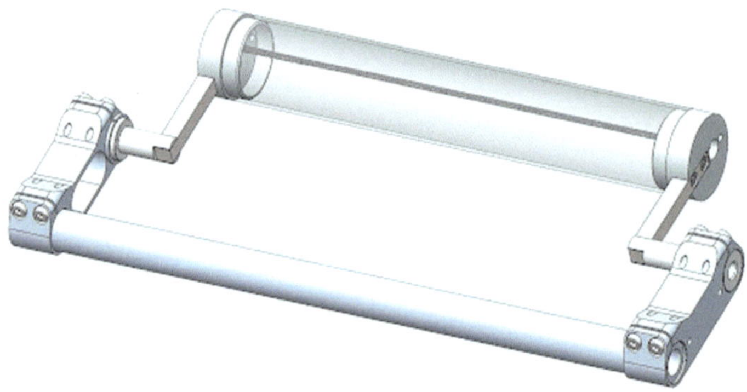

Fig. 4.12 Version 7 tuning arm

Table 4.2 Cavity version 7 empty configurations Q factors

Axles	Screws	Arms	Frequency (GHz)	Q_U
Removed	Removed	Removed	1.793	18,800
PEEK	Removed	Removed	1.793	18,770
PEEK	PEEK	PEEK	1.78	15,100
PEEK	PEEK	Aluminum	1.78	17,200

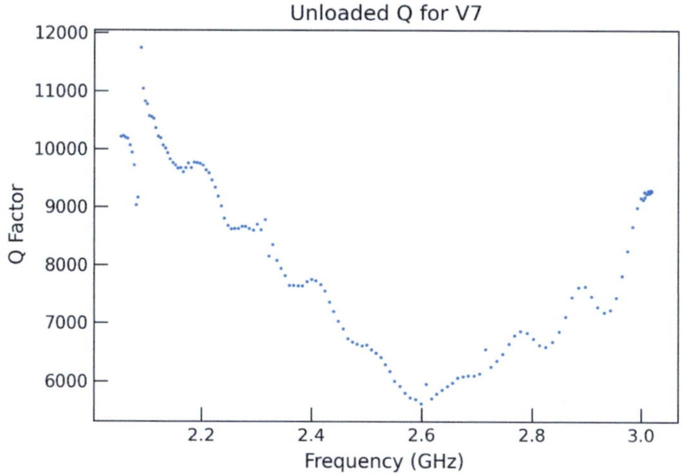

Fig. 4.13 Unloaded Q for the version 7 cavity

out of the cavity. The version 8 cavity was designed with the thought of preventing this leakage in mind.

4.4.4 Impedance Filters

The use of impedance filters was proposed to reduce the leakage of power out of the axle of the tuning rod. Given the final frequency range will be 2 GHz to 4 GHz, either a low-pass filter with a cutoff of approximately 2 GHz or a high-pass filter with a cutoff of 4 GHz could be employed. The design chosen was a stepped impedance coaxial low-pass filter [1]. The basic design is alternating components with low and high impedance. The design employed for testing was a 7th order impedance filter, shown in Fig. 4.14.

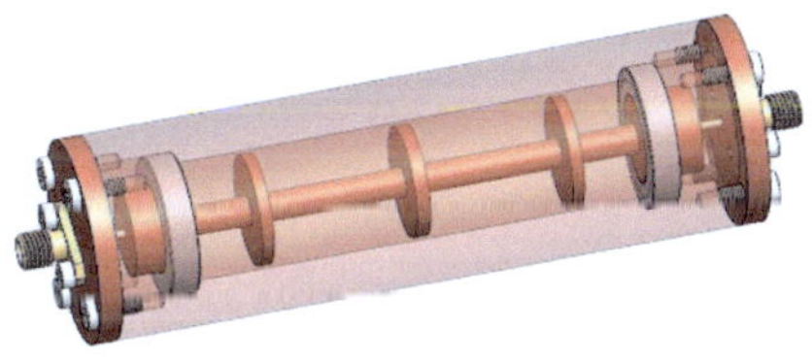

(a) Impedance filter v1.

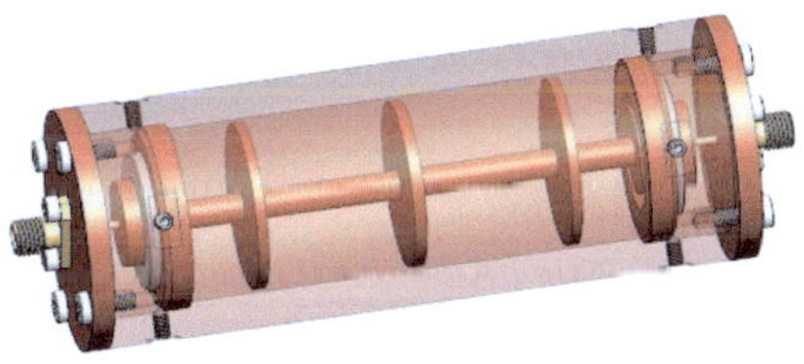

(b) Impedance filter v2.

Fig. 4.14 Impedance filter diagrams

Fig. 4.15 Prototype of the v2 impedance filter

A calculator was employed to compute the parameters of each component [2]. The first prototype, v1, was calculated using a cutoff of 2 GHz, however it was realized the width of the transition band needed to be taken into account. A second impedance filter, v2, was designed with a cutoff of approximately 1.1 GHz. A prototype of the v2 impedance filter is shown in Fig. 4.15.

To evaluate the performance of the impedance filters, an S21 measurement was done to record the level of attenuation across 2 to 3 GHz. The S21 spectra of these measurements are shown in Fig. 4.16. As seen, there is a significant drop in the power

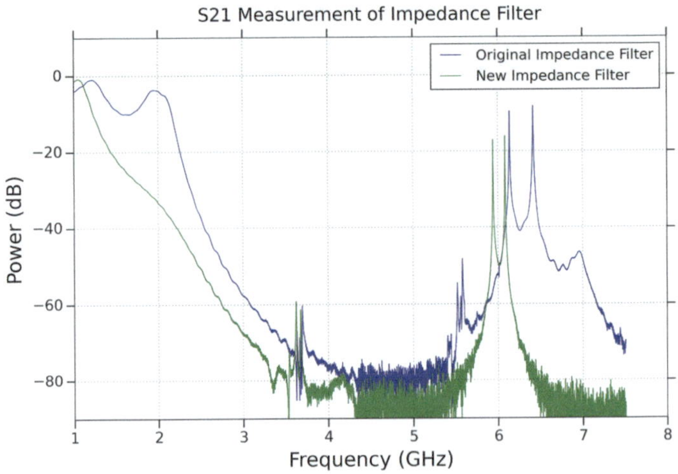

Fig. 4.16 Comparison of the v1 and v2 impedance filters

of the signal transmitted by the filters between 2 and 3 GHz. This means when the filters are attached to the cavity, power that previously would have escaped the cavity gets reflected back in. Both impedance filter versions supported an unexpected pair of modes near 6 GHz, however given these were outside the frequency range of interest, the modes were ignored.

4.4.5 Cavity Version 8

The version 8 cavity had the same dimensions as the version 7 cavity, adding turrets to both sides of the tuning mechanism. An improvement in Q_U was expected with even just the turrets in place and no impedance filters. The additional length of the turrets was thought to reflect more power back into the cavity. A diagram of version 8 is shown in Fig. 4.17.

Three sweeps were done to calculate Q_U using a metal arm, PEEk arm, and teflon arm. The results are shown in Fig. 4.18. Like version 7, the metal arms had the best performance and were used for subsequent tests. A noticeable improvement from the version 7 Q_U measurements was observed.

The turrets were used as a housing for the filters, shown in Fig. 4.19. The first measurement using the v1 filter performed worse than with just the turrets, shown in Fig. 4.20.

It was assumed there was unwanted contact between the impedance filter and its housing, resulting in the poor performance. A 0.002" plastic spacer was also inserted around the axle it prevent contact between it and the end plate, shown in Fig. 4.21.

After testing this configuration, it was determined the metal arm holding the tuning rod could be made thinner by 2/3, which was done and another cavity sweep was ran. These results are shown in Fig. 4.22. In this configuration, new TE modes began to couple very strongly to the TM_{010} mode, resulting in a large dip in Q_U.

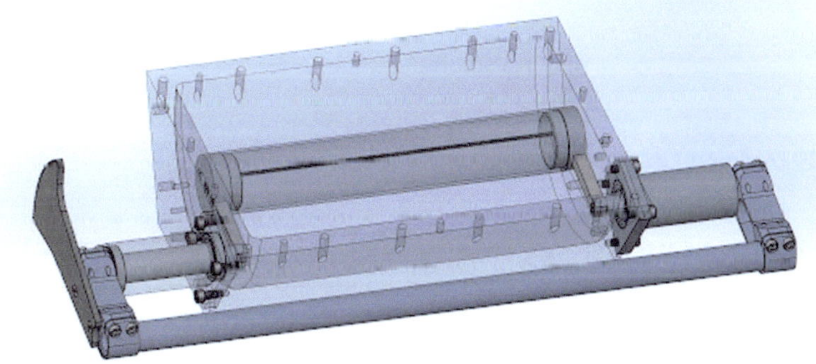

Fig. 4.17 Diagram of the version 8 cavity

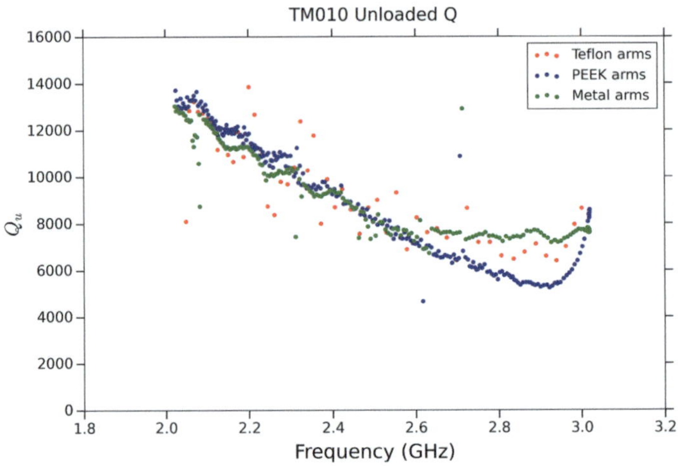

Fig. 4.18 Comparison of the unloaded Q between version 8 with PEEk, teflon, and metal arms

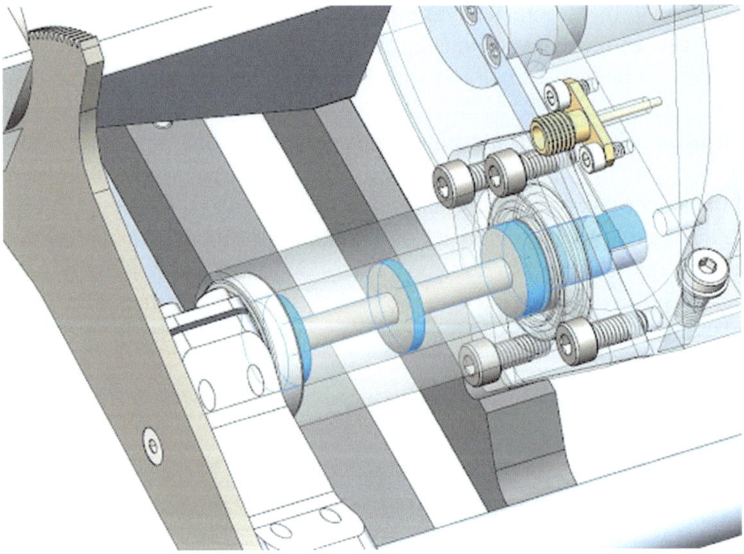

Fig. 4.19 Diagram of the impedance filter attached to the cavity

An attempt to fix this was done by removing even more material from the impedance filter, increasing its clearance relative to the turret, and removing the plastic spacer. A comparison of this test to the previous two is shown in Fig. 4.23. This solution significantly improved Q_U, to a level nearly acceptable as a final design.

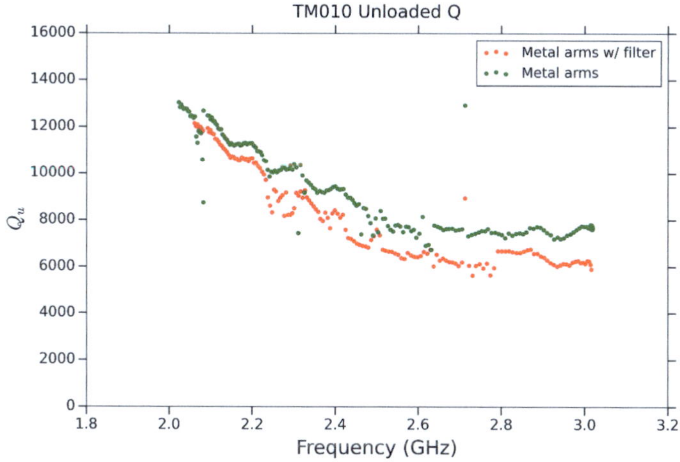

Fig. 4.20 Comparison of the unloaded Q between version 8 with metal arms and no filter versus the first impedance filter

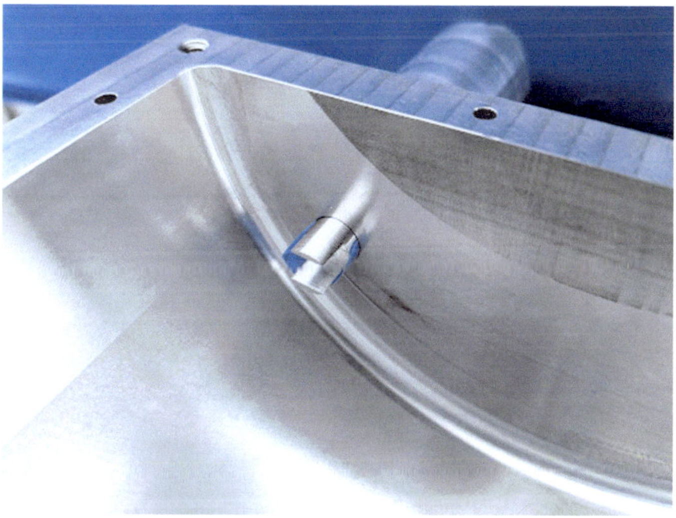

Fig. 4.21 Plastic spacer around the axle of the version 8 cavity

Next, the v2 impedance filter was tested. A comparison of its performance to the v1 filter is shown in Fig. 4.24. The v2 impedance filter performed moderately better. This indicated a final design had been approximately reached, aside from fine tuning. A comparison of this cavity configuration to the previous versions discussed is shown in Fig. 4.25, where it can be seen that the V8 cavity with the v2 impedance filter yielded the best performance.

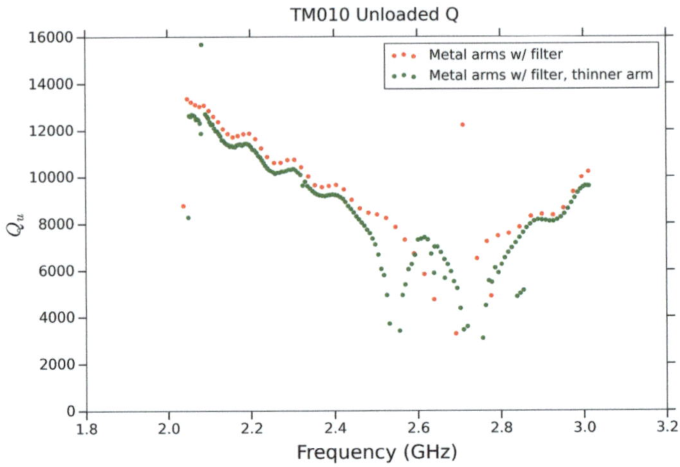

Fig. 4.22 Unloaded quality factor for version 8 after removing material from the impedance filter, adding a plastic spacer, then thinning the metal arm

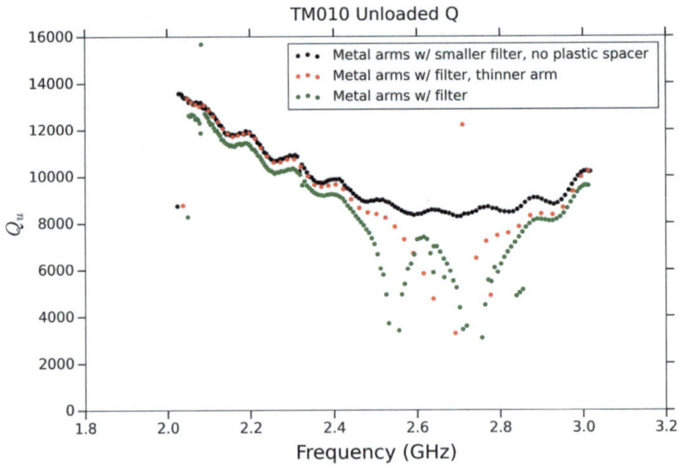

Fig. 4.23 Comparison of the unloaded Q after removing the plastic spacer and removing more material from the filter

4.5 ADMX EFR Testing Conclusions

The optimization of Q_U for the EFR cavity was a tedious process. At several stages, proposed solutions resulted in worse performance. However, the solution was not necessarily incorrect, but rather just implemented poorly, as was the case for the initial impedance filter. The work done was able to make two key conclusions about

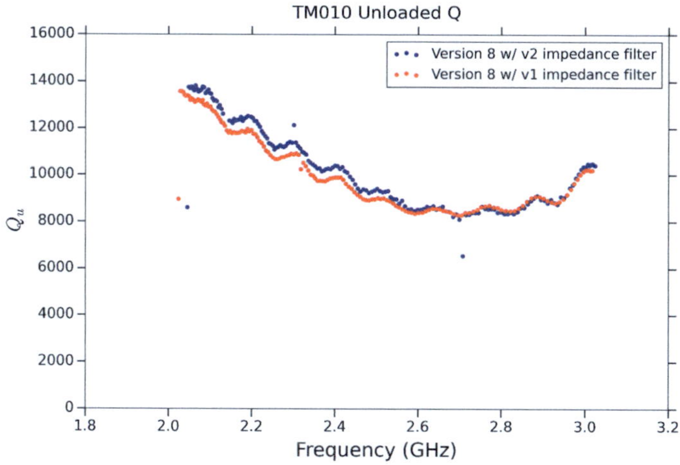

Fig. 4.24 Comparison of the v1 and v2 impedance filter performances

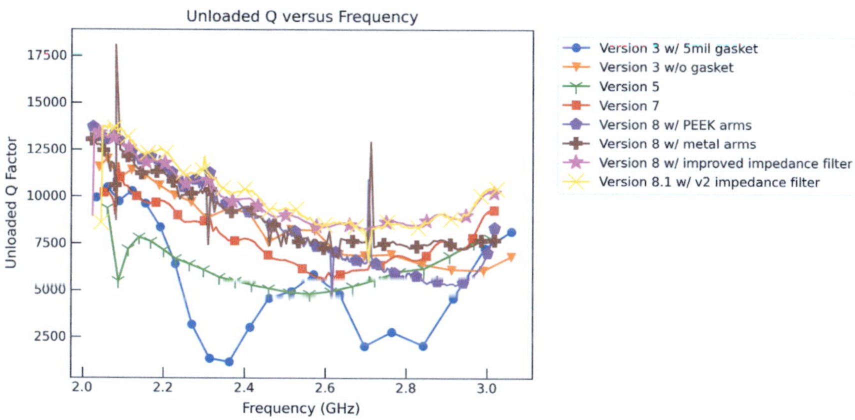

Fig. 4.25 Comparison of the unloaded Q across all major versions

resonant cavity designs. First, a slotted cavity can result in significant power loss, leading to a reduced Q_U, so an internal tuning mechanism should be preferred. Second, impedance filters can be used to prevent the coupling of power out of a resonant cavity, reducing another source of power loss. These filters can be designed for a large range of frequencies and thus can be utilized by many experiments. Finally, a near final design, with respect to optimizing Q_U, was achieved.

References

1. S. Das, D.S.K. Chowdhury, Design simulation and fabrication of stepped impedance microstripline low pass filter for S-band application using IE3D and MATLAB (2012)
2. A.C.F. Changpuak aka, Stepped impedance coaxial lowpass filter designer. https://www.changpuak.ch/electronics/Stepped_Impedance_Lowpass_Coax.php

Chapter 5
LC Circuit Axion Search

Abstract The standard cavity haloscope search becomes nearly impossible for axions with a mass below approximately 1 μeV, since the cavity would need a diameter greater than a few meters, which is impractical. Instead, an LC resonator can be employed. This chapter discusses the development of the ADMX Superconducting LC circuit Investigating Cold dark matter (SLIC) experiment, an LC resonator searching for axions with a mass between 1.861×10^{-7} eV to 2.192×10^{-7} eV. This chapter discusses the theory, expected signal strength, challenges, and search strategy.

5.1 Low-Frequency Searches

The primary motivation for low-mass axions is their inclusion in GUTs. For example, an $SU(5) \times U(1)_{PQ}$ model predicts an axion with a mass on the order of a neV [1]. To perform a cavity haloscope search, the cavity's diameter should be approximately equal to the axion's Compton wavelength. An axion with a mass of 0.2 μeV has a Compton wavelength of about 6 m. A cavity of this size could not fit inside the bore of a typical solenoid magnet and constructing this cavity and a custom magnet would be extremely costly. One solution is to use a different type of resonator. A proposal by Sikivie, Sullivan, and Tanner suggested the use of an LC circuit, which could probe an axion mass range of approximately 10^{-9} eV to 10^{-7} eV [2]. Crisosto and collaborators developed the ADMX Superconducting LC Circuit Investigating Cold Axions (ADMX SLIC) Experiment based on this concept [3]. This search placed strong limits on axions with a mass of approximately 1.8×10^{-7} eV, demonstrating the capability of this search type.

A continuation of ADMX SLIC aims to perform a second data taking run, tuning over approximately 45 MHz to 53 MHz. This chapter includes a discussion of the theory behind SLIC, the current progress on the project, and future plans.

A. Hipp, *Searching High and Low for the Axion*,
Springer Theses, https://doi.org/10.1007/978-3-032-02276-9_5

5.2 LC Circuit Dynamics

An LC circuit is one of the simplest circuit designs. The resonant frequency of an ideal LC circuit is given by

$$\omega_0 = \frac{1}{\sqrt{LC}}, \tag{5.1}$$

where C is the capacitance and L is the inductance of the circuit. A real LC circuit will have some resistance, R, in which case the circuit behaves according to

$$\ddot{I}(t) + \frac{R}{L}\dot{I}(t) + \frac{I(t)}{LC} = 0, \tag{5.2}$$

which describes a damped oscillator, with damping factor $\zeta = \frac{R}{2}\sqrt{\frac{C}{L}}$. The over-damped and critically damped cases are of no interest, as the circuit does not oscillate. The underdamped case has $\zeta < 1$, with the solution

$$I(t) = I_0 e^{-\frac{Rt}{2L}} sin(\omega t + \phi), \tag{5.3}$$

where $\omega = \omega_0\sqrt{1-\zeta^2}$ and ϕ is some phase shift. The quality factor, defined in Appendix B, describes the energy loss of the circuit and is given by

$$Q = \frac{1}{2\zeta} = \frac{\omega_0 L}{R}. \tag{5.4}$$

5.3 Axion Detection with an LC Circuit

5.3.1 *Detecting the Axion Signal with an LC Circuit*

The LC circuit used by ADMX SLIC consisted of a parallel plate capacitor and a single rectangular loop of wire. The LC circuit was placed inside a solenoid magnet. The process of an axion passing through a solenoid magnet is described in Appendix A. Figure 5.1 is a diagram of the LC circuit in a solenoid magnet.

Using Eq. A.9, it is straightforward to find

$$\mathbf{B}_a(r) = -\frac{1}{2} g B_0 \frac{\partial a}{\partial t} r\hat{\phi}, \tag{5.5}$$

where r is the radius, in cylindrical coordinates. The rectangular loop of wire captures the axion magnetic field flux, Φ_a, driving a current in the circuit. Computing the magnetic flux in the usual way yields

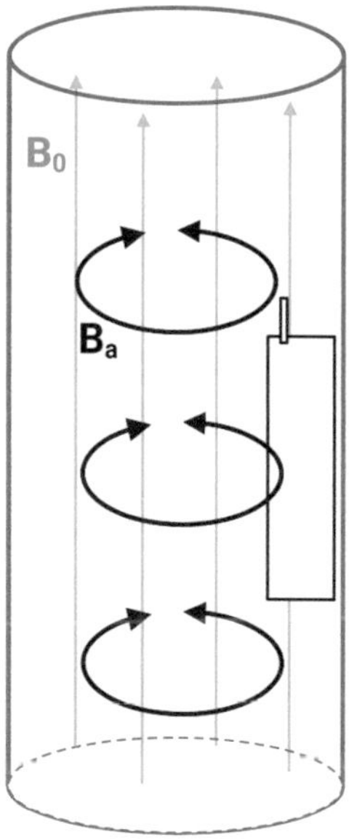

Fig. 5.1 Diagram of LC circuit in a solenoid magnet, driven by an axion magnetic field. The magnetic field oscillates, changing direction, shown by the arrows in both directions

$$\Phi_a = -V_m g B_0 \frac{\partial a}{\partial t}, \tag{5.6}$$

where $V_m = \frac{1}{4}hw^2$ is the effective volume of the main loop, with h and w being the height and width of the loop of wire, respectively. Labeling the inductance of the rectangular loop of wire by L, the current in the circuit is given by

$$I_a = \frac{Q}{L} V_m g B_0 \frac{\partial a}{\partial t}, \tag{5.7}$$

where the quality factor, Q, enhances the current.

5.3.2 Expected Axion Power of the LC Circuit

To read out the power, a pick-up loop was inductively coupled to the LC circuit, described in Appendix C. The effective axion magnetic field induces a current in the LC circuit, which then produces its own magnetic field, whose flux is captured by the pick-up loop. The flux captured by the readout pick-up loop depends on the mutual inductance, M, between the main loop and pick-up and is given by

$$\Phi_p = M I_a = \kappa \sqrt{L_1 L_2} I_a, \tag{5.8}$$

where $M = \alpha\sqrt{L_1 L_2}$, with κ a coupling constant, L_1 the inductance of the main loop, and L_2 the inductance of the pick-up loop. The induced voltage in the pick-up loop is then found to be

$$\begin{aligned} V_p &= -N\frac{d\Phi_\text{p}}{dt} \\ &= -N\kappa\sqrt{L_1 L_2}\dot{I}_a \\ &= -N\kappa\omega\sqrt{L_1 L_2}I_a, \end{aligned} \tag{5.9}$$

where N is the number of turns in the pick-up loop. The power is then given by

$$\begin{aligned} P &= \frac{V_\text{p}^2}{\text{R}} \\ &= \frac{N^2\kappa^2\omega^2(L_1 L_2)I_a^2}{R}. \end{aligned} \tag{5.10}$$

Substituting in the axion current from Eq. 5.7 and using $\dot{a}^2 = 2\rho_a$, the signal power is found to be

$$\begin{aligned} P &= 2N^2\kappa^2\omega^2\frac{L_1 L_2}{R}\left(\frac{V_m^2 Q^2}{L_1^2}g_{a\gamma\gamma}^2 B_0^2\rho_a\right) \\ &= 2N^2\kappa^2\omega^2\frac{Q^2 L_2}{L_1 R}\left(V_m^2 g_{a\gamma\gamma}^2 B_0^2\rho_a\right). \end{aligned} \tag{5.11}$$

As discussed in Sect. 2.3.1, the quality factor of the axion flow would replace Q, in the event Q is larger. Recall several of the computations performed were done using natural units. Performing dimensional analysis, it is easy to see the above expression can be converted to SI units by multiplying by a factor of $\hbar c$.

As was done in Sect. 2.3.6, normalizing the power to the experimental parameters, in SI units, is more useful. Substituting in Eq. A.7 for $g_{a\gamma\gamma}$, using $m_a = 2\pi\hbar f$, and $\omega = 2\pi f$ yields

$$P = 32\pi^2\hbar^3 c\kappa^2 N^2\alpha^2\frac{Q^2 L_2}{L_1 R}f^4 V_m^2 B_0^2\rho_a g_\gamma^2\left(\frac{1}{6\times 10^{15}\ \text{eV}^2}\right). \tag{5.12}$$

Using the expected experimental parameters of $Q = 4,500$, the values of L_1 and L_2 from Appendix C, $\kappa = 0.5$, $V_m = 0.00046\ \text{m}^3$, $B_0 = 4.5\text{T}$, $N = 5$, $g_\gamma = 0.36$ (DFSZ), $\rho_a = 0.45\ \text{GeVm}^{-3}$, $R =\ \Omega$, and $f = 50$ MHz yields

$$\begin{aligned} P = 1.13\times 10^{-26}\ \text{W}\left(\frac{N}{5}\right)^2\left(\frac{\kappa}{0.5}\right)^2\left(\frac{Q}{4,500}\right)^2\left(\frac{L_2}{1.4\ \mu\text{H}}\right)\left(\frac{0.968\ \mu\text{H}}{L_1}\right)\left(\frac{50\ \Omega}{R}\right) \\ \left(\frac{g_\gamma}{0.36}\right)\left(\frac{f}{50\ \text{MHz}}\right)^4\left(\frac{\rho_a}{0.45\ \text{GeVcm}^{-3}}\right). \end{aligned} \tag{5.13}$$

Comparing to Eq. 2.22, the expected power for the LC circuit is approximately 6,000 times smaller than the ADMX Run 1c expected power.

5.4 Inherent Losses

A physical circuit has several unavoidable sources of losses, such as skin effects and dielectric losses. However, the losses can be reduced by choosing the proper materials for conductors and support structures. The following sections discuss these losses and the impact on material selection.

5.4.1 *Skin Effects*

For conductors carrying an alternating current, the skin effect must be taken into consideration. This is characterized by a frequency skin depth given by

$$\delta(\omega) = \sqrt{\frac{2}{\sigma \mu \omega}}, \tag{5.14}$$

where ω is the frequency, σ is the conductor's conductivity, and μ is the conductor's permeability. At low frequency, the resulting surface impedance is real, given by

$$Z = \frac{1}{\sigma \delta}. \tag{5.15}$$

For superconductors, the equivalent characterization is the London penetration depth, given by

$$\lambda_L = \sqrt{\frac{m}{\mu_0 n_s e^2}}, \tag{5.16}$$

where m and e are the mass and charge of the electron, respectively, μ_0 is the permeability of free space, and n_s is the super carrier density. A second characteristic length scale of superconductors is the coherence length, which describes the distance over which the super carrier density can change, given by

$$\xi_0 = \frac{2}{\pi} \frac{\hbar v_F}{2\sqrt{n_s}}, \tag{5.17}$$

where $\hbar$ is the reduced Planck constant, and v_F is the Fermi velocity [4]. Notice the London penetration depth is frequency independent. For room temperature testing, S_{21} and quality factor measurements were made using a copper pick-up loop, which has a skin depth of $\delta \approx 10$ µm. The final LC circuit was constructed using NbTi capacitor plate and 0.25 mm diameter NbTi wire. The London penetration depth and coherence length of NbTi has been estimated to be $\lambda_L \sim 100$ nm and $\xi_0 \sim 20$ nm [4]. The small λ_L justified the use of the 0.25 mm diameter wire.

5.4.2 Dielectric Losses

The dielectric slab for tuning and the support structure for the LC circuit introduce additional losses. In general, the permittivity of a material is a complex function of angular frequency

$$\varepsilon(\omega) = \varepsilon_1(\omega) - i\varepsilon_2(\omega). \tag{5.18}$$

The loss tangent of a material is given by $\tan\delta = \frac{\varepsilon_2}{\varepsilon_1}$. The dissipation factor, or quality factor, of a material is the inverse of the loss tangent

$$Q = \mathrm{DF} = \frac{1}{\tan\delta}. \tag{5.19}$$

The dielectric slab used was made of alumina, which has $\tan\delta = 0.0002$. A study comparing the performance of teflon (PTFE) and Formvar insulation on the NbTi wire, as well as teflon versus PEEK for the support structure [5]. The study found teflon performed better in both cases, so it was employed in the final LC circuit design.

5.4.3 Radiation Resistance

An antenna has the same radiation resistance when receiving and transmitting. For a small loop, satisfying

$$C \leq \frac{\lambda}{5}, \tag{5.20}$$

where C is the loops circumference and λ is wavelength, the radiation resistance is given by [6]

$$R_r = 320\pi^4 \left(\frac{A}{\lambda^2}\right) \Omega, \tag{5.21}$$

where A is the area of the loop. The main pick-up loop in the LC circuit is the one-loop inductor. With dimensions of 3" × 12.5", the circumference is $C \approx 0.8128$ m. The largest frequency of the LC circuit is 53.44 MHz, corresponding to a wavelength of approximately 5.61 m, so Eq. 5.20 is satisfied. The radiation resistance is then $R_r \approx 0.0251\ \Omega$.

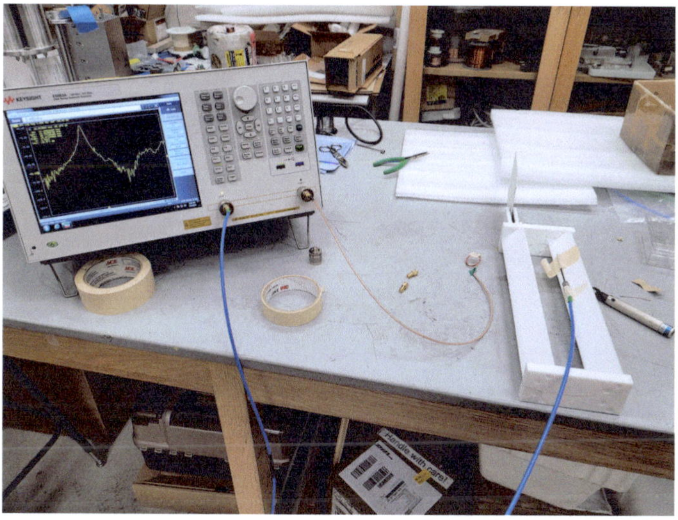

Fig. 5.2 Warm testing of the LC circuit

5.5 Warm Testing

The continuation of the SLIC experiment utilized the LC circuit originally used in the first data taking run. Room temperature testing of the circuit was done to confirm its tunability. The resonator was tested warm, using a pin antenna to excite the circuit and a copper pick-up loop to detect the resonance. The testing set-up is shown in Fig. 5.2.

Three measurements were made by inserting an alumina slab, with dimensions 2" × 2" × 0.06", half-way and fully into the parallel plate capacity, and with no dielectric. Figure 5.3 shows the S21 spectra captured for each measurement. The resonant frequencies measured were approximately 53 MHz, 48 MHz, and 45 MHz for no alumina, half-way inserted, and fully inserted, respectively. These results matched well with the predicted tuning range, discussed in Appendix C.

5.6 SLIC Electronics

5.6.1 Linear Actuator

The first run of SLIC utilized a rotary actuator to tune the circuit. The alumina slab was rotated into the parallel plate capacitor. However, unknown issues with the tuning mechanism restricted the tuning range to 10s of kHz. This project will instead use a linear actuator to simply raise and lower the alumina slab into the capacitor. A JPE

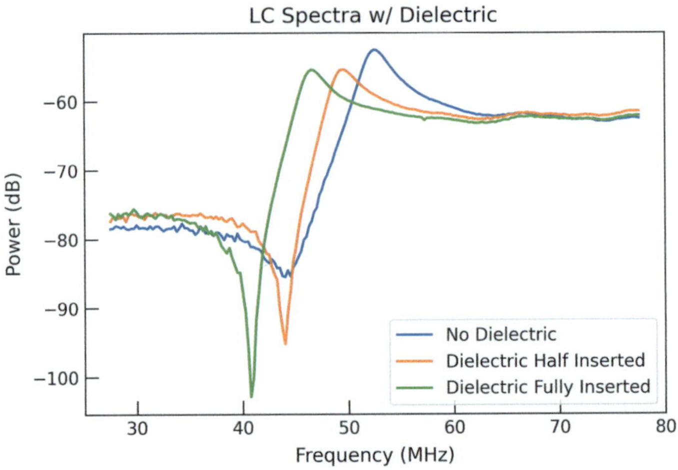

Fig. 5.3 S21 spectrum of the LC circuit with an alumina slab outside, half-way inserted, and fully inserted

Cryo Linear Drive (CLD) was chosen as the actuator[7]. The JPE CLD has a stroke of 51.5 mm, a large enough range of motion to fully insert the alumina slab. The JPE CLD is rated to have a minimum step size of 100 nm, so for the expected 8.5 MHz tuning range, the minimum tuning step 16 Hz enabling fine tuning to be performed. The JPE CLD is non-magnetic and can operate at cryogenic temperatures, making it a suitable choice. A python script was written to control the linear actuator remotely.

5.6.2 *SLIC DAQ*

The data taking process will effectively include three steps; run the linear actuator to tune the circuit, perform a quality factor and reflection measurement, and collect data. The electronics used to accomplish this were a Keysight E5063A VNA, HP8656A signal generator, and SR 760 spectrum analyzer. A diagram of the SLIC DAQ is shown in Fig. 5.4. As discussed in Appendix C, the LC circuit is inductively coupled to a five-turn pick-up loop, which is connected to a Stahl HDC-50 amplifier. The signal from the Stahl HDC-50, at cryogenic temperature, is run to a Stahl A3-7 for biasing and room temperature post-amplification using a tinned copper braid PTFE insulated coaxial cable [8]. From there, the signal is directed to a ZX80-DR230-S+ RF switch [9], where it will then be directed to either the Keysight E5063A VNA or to the RF port of a ZX05-1-S+ mixer [10]. The state of the RF switch is controlled using an NI USB-6501 [11]. An HP8656A signal generator is connected to the LO port of the mixer and the SR 760 Spectrum Analyzer to the IF port.

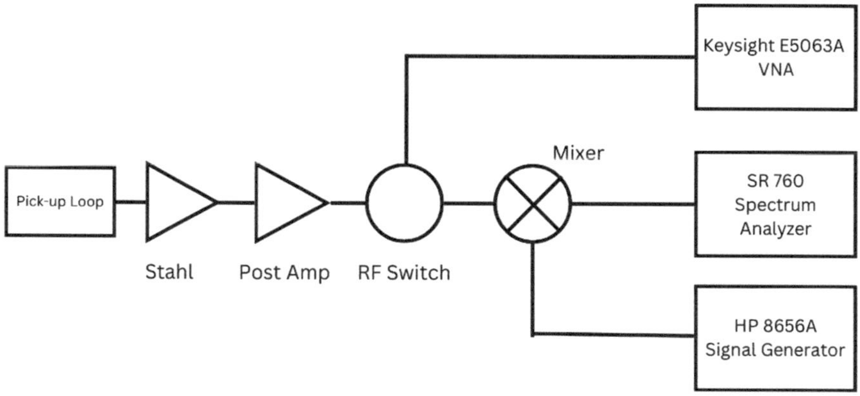

Fig. 5.4 A diagram of the ADMX SLIC DAQ

5.6.3 SLIC Data Taking

The data taking process will be automated using scripts written in Python to control the electronics. The process will the steps

1. Tune the circuit by running the linear actuator
2. Direct the RF switch to the Keysight VNA
3. Identify the resonant frequency
4. Perform Q and reflection measurements
5. Set the local oscillator frequency
6. Set RF switch to the mixer
7. Average power for T seconds with SR 760 SA.

The data collection time, T, is initially planned to be 100 s. To perform a Q measurement, the resonant frequency, f_0 is found using the peak-finder function of the Keysight E5063A VNA. In step 5, the local oscillator is then set to $f_0 + 50$ kHz, so the RF signal is mixed down to 50 kHz and sent to the spectrum analyzer. The S_{21}, S_{11}, and S_{22} spectra from the Keysight VNA are written to an hdf5 file and the power spectrum from the SR 760 are written to a CSV file.

5.7 SLIC Next Steps

The current stage of the experiment is the electronics, control scripts, and circuit design, have been completed. The next step is to prepare the helium system and magnet for data taking, and to run cabling from the room temperature stage to the cold stage for the linear actuator. The magnet is a Cryomagnetics superconducting (NbTi) solenoid, rated to have an average magnetic field strength of 7.5 T. However, the first data taking run saw the magnetic field drop to 4.5 T, due to power supply instabilities. There are plans to fix this, so a 7 T field is had for the entire data taking run.

5.8 SLIC Expectations

Progress has been made in the recommissioning of the ADMX SLIC experiment. Once the testing of the cryogenics and magnet is completed, a data taking run will be done. The expectation is appreciable limits will be placed on the axion parameter space. The first ADMX SLIC data run ruled an axion-photon-photon coupling of approximately 3×10^{-13} GeV^{-1} across a frequency range of about 80 kHz [3]. A similar sensitivity is expected to be reached across an 8.4 MHz tuning range, corresponding to an axion mass range of 1.861×10^{-7} eV to 2.192×10^{-7} eV. As mentioned in Appendix C, the use of an alumina slab with a 0.1" thickness would increase the tuning range to approximately 38 to 53 MHz, corresponding to an axion mass range of 1.572×10^{-7} eV to 2.192×10^{-7} eV.

References

1. A. Ernst, L. Di Luzio, A. Ringwald, C. Tamarit, in *Proceedings of Corfu Summer Institute 2018 "School and Workshops on Elementary Particle Physics and Gravity"—PoS(CORFU2018)* (Sissa Medialab, Corfu, Greece, 2019) p. 054. https://doi.org/10.22323/1.347.0054
2. P. Sikivie, N. Sullivan, D. Tanner, Phys. Rev. Lett. **112**, 131301 (2014). https://doi.org/10.1103/PhysRevLett.112.131301
3. N. Crisosto et al., Phys. Rev. Lett. **124**, 241101 (2020). https://doi.org/10.1103/PhysRevLett.124.241101
4. X. Xi, J. Hwang, C. Martin, D.B. Tanner, G.L. Carr, Phys. Rev. Lett. **105**, 257006 (2010). https://doi.org/10.1103/PhysRevLett.105.257006
5. C.M. Crisosto, Searching for low mass axions with an LC circuit. Ph.D. thesis, University of Florida, 2018
6. N.K. Nikolova, Lecture notes antenna engineering. https://www.ece.mcmaster.ca/faculty/nikolova/antenna_dload/current_lectures/LectureNotesAntennas_Nikolova.pdf
7. https://www.jpe-innovations.com/cryo-uhv-products/cryo-linear-drive/
8. https://www.stahl-electronics.com/bilder/datasheet_a3_a7_v2_5.pdf
9. https://www.minicircuits.com/pdfs/zx80-dr230+.pdf
10. https://www.pasternack.com/images/productpdf/pe-sr405fl.pdf
11. https://www.ni.com/docs/en-us/bundle/usb-6501-specs/page/specs.html

Chapter 6
Future Work

Abstract Much of the work in this thesis can and is currently being built upon. ADMX will continue to perform multi-res analyses for future data taking runs, ADMX-EFR is still under development, and ADMX SLIC is near a data taking run. This chapter introduces possible future endeavors for each of these projects.

6.1 ADMX Multi-Res Continued

The multi-resolution analysis channel could be improved in several ways. First, the analysis for run 1c was performed offline, once data taking had been completed. If a candidate had been found, rescanning it would have been very difficult, as it would have required a cool down and magnet ramp, which is a time consuming and costly procedure. A solution to this would be to either implement a script that analyzes and identifies triggers during live data taking, or to analyze the data offline, but at the end of each nibble. This would allow for the rescan of multi-resolution triggers in a timely manner.

Another improvement would be to perform a detailed study of the SNR versus frequency resolution. The frequency resolutions of 20 mHz, 40 mHz, 200 mHz, and 1000 mHz were chosen somewhat arbitrarily. A study to determine what resolutions from 20 mHz to 1 Hz would provide the best overlapping coverage with respect to SNR would make the probability of detecting a signal more agnostic to the signal dynamics.

Lastly, the trigger thresholds were chosen based on the probability distributions of the background. However, there could be clever ways to lower the threshold further, without identifying too many triggers. For example, it takes several tuning steps for a signal to pass through the cavity bandwidth. Thus, a cut demanding that for a trigger within the bandwidth of one digitization, there must be another trigger within a small frequency range, shifted by the tuning step, in the proceeding digitization. clever analysis techniques such as this could potentially lower the threshold by a few standard deviations.

A. Hipp, *Searching High and Low for the Axion*,
Springer Theses, https://doi.org/10.1007/978-3-032-02276-9_6

6.2 Future of ADMX EFR

The work discussed in Chap. 4 was the first stage in designing a cavity with an optimized Q_U, so there is plenty of work that can still be done. First, the prototypes tested had a length of approximately 20 cm, while the cavities that will be used for data taking will have a length of 1 m. A full scale prototype should be built and tested. The TE modes that will be present during data taking would be revealed. These could potentially interfere with the TM_{010} mode strongly across large frequency ranges, requiring design changes. Additionally, the final cavities will be made of copper, so building and testing a copper or copper-plated prototype would yield a better estimate for the quality factor. Lastly, a cold test of the prototype would also yield a better estimate for expected quality factor.

6.3 Future of ADMX SLIC

Based on current progress, the ADMX EFR cryogenics and magnet may operational before the EFR experiment is ready to be ran. There may be the opportunity to perform an ADMX SLIC run using the EFR infrastructure. The EFR magnet has a magnetic field strength of 9 T, compared to the SLIC magnet with a field strength of approximately 4.5 T. This would yield a factor of four increase in the axion power. Re-configuring the dimensions of the LC circuit to make better use of the additional space inside the magnet, an LC search using the EFR infrastructure could very well lead to QCD axion sensitivities being reached.

Chapter 7
Conclusion

Abstract This chapter provides concluding remarks about the impact of the work discussed.

The search for axion dark matter requires a broad range of techniques, as demonstrated by the work discussed. A multi-resolution analysis can make an axion search sensitive to any dark matter halo model. The power of this analysis technique was shown in this work through its application to the ADMX run 1c dataset. Assuming the caustic ring model, the analysis confidently ruled out DFSZ axions. Assuming the DFSZ axion, the four expected flows of the caustic ring model were ruled out, along with any nonvirialized flow with a density greater than 0.1 GeV/cm^{-3}. The work done for the development of an ADMX EFR cavity yielded promising results. The optimized quality factor increases the scan rate for ADMX EFR and the stepped impedance tests demonstrated their usefulness. Recommissioning of SLIC has neared its final stages, with a data taking run close to being done. Similar limits to the first SLIC data taking run are expected to be extended to a tuning range of approximately 8 MHz. Overall, the work presented made significant contributions toward the axion search effort, hopefully making its discovery ever nearer.

A. Hipp, *Searching High and Low for the Axion*,
Springer Theses, https://doi.org/10.1007/978-3-032-02276-9_7

Appendix A
Axion Equations of Motion

The Lagrangian density for the axion-photon interaction is given by [1]

$$\mathcal{L} = -\frac{1}{4}F_{\mu\nu}F^{\mu\nu} + \frac{e^2N}{12\pi^2}\frac{a}{v}F_{\mu\nu}\tilde{F}^{\mu\nu} + \frac{1}{2}\partial_\mu a\partial^\mu a - \frac{1}{2}m_a^2 a^2\left[1 + \mathcal{O}(\frac{a^2}{v^2})\right], \quad \text{(A.1)}$$

where $F_{\mu\nu}$ is the electromagnetic field strength tensor and $\tilde{F}^{\mu\nu} = \frac{1}{2}\epsilon^{\mu\nu\alpha\beta}F_{\alpha\beta}$ is its Hodge dual, a is the axion field, m_a is the axion mass, v is the Peccei-Quinn symmetry breaking scale, and N is the number of vacua in the axion model. The equations of motion are given by

$$\nabla \times \vec{B} - \frac{\partial \vec{E}}{\partial t} = g[\vec{E} \times \nabla a - \vec{B}\frac{\partial a}{\partial t}] \quad \text{(A.2)}$$

$$\nabla \cdot \vec{E} = g\vec{B}\ \nabla a \quad \text{(A.3)}$$

$$\Box a = ga\vec{E} \cdot \vec{B} - m_a^2 a. \quad \text{(A.4)}$$

A model-dependent coupling constant g has been introduced to simplify coefficients in Eq. A.1. This coupling is defined as [2]

$$g_{a\gamma\gamma} = g_\gamma \frac{\alpha}{\pi}\frac{1}{f_a}, \quad \text{(A.5)}$$

where $\alpha = \frac{e^2}{4\pi}$ is the fine structure constant and g_γ is a model dependent coupling, given by

$$g_\gamma = \frac{1}{2}\left[\frac{N_e}{N} - \frac{5}{3}\frac{m_d - m_u}{m_d + m_u}\right], \quad \text{(A.6)}$$

where N is the color anomaly of the PQ charge, N_e is the electromagnetic anomaly of the PQ charge, m_u is the up quark mass, and m_d is the down quark mass. Solving Eq. 2.5 for f_a and substituting the result into Eq. A.5

A. Hipp, *Searching High and Low for the Axion*,
Springer Theses, https://doi.org/10.1007/978-3-032-02276-9

$$g_{a\gamma\gamma} \simeq g_\gamma \frac{\alpha}{\pi} \frac{m_a}{6\mu\text{eV} \times 10^{12}\ \text{GeV}}. \tag{A.7}$$

For the two benchmark models, KSVZ uses $N = 1$ and $N_e = 0$ predicting $g_\gamma \simeq -0.97$, and DFSZ uses $N = 6$ and $N_e = 16$ predicting $g_\gamma \simeq 0.36$. Experiments based on the axion-photon-photon interaction aim to be sensitive to the couplings predicted by KSVZ and DFSZ. As shown in Fig. 2.8, the limits are always set on the absolute value of the coupling, so the sign disparity between DFSZ and KSVZ is irrelevant.

In the first line of Eq. A.2, consider the case of an externally applied static magnetic field, $\vec{B}_0$. The axion field acts an effective current, given by

$$\vec{j}_a = -g\vec{B}_0 \frac{\partial a}{\partial t}, \tag{A.8}$$

which, according to Ampere's law, results in an axion sourced magnetic field given by

$$\nabla \times \vec{B}_a = \vec{j}_a. \tag{A.9}$$

The axion induced current is frequency dependent, $\vec{j}_a$, oscillates at a frequency set by the energy of the axion, given by

$$\omega_a = m_a(1 + \frac{1}{2}\vec{v}_a \cdot \vec{v}_a), \tag{A.10}$$

where $\vec{v}_a$ is the axion velocity. Various dark matter halo models predict different distributions for the axion velocity and thus the frequency spread of the resulting signal.

References

1. P. Sikivie, Phys. Rev. Lett. **51**, 1415 (1983)
2. P. Sikivie, Rev. Mod. Phys. **93**, 015004 (2021)

Appendix B
Two-Port Cavity Measurements

Vector Network Analyzers (VNA) can be used to measure the unloaded quality factor of resonant cavities [1]. The most common setup is a two-port measurement, where one port is transmitting power into the cavity and the second is measuring the power out of the cavity [2]. VNAs measure the four S-parameters, labeled S_{BA}, which gives the power transmitted from port A to port B, where A and B correspond to port 1 or port 2. For a two port cavity, S_{12} and S_{21} are equivalent and give the cavity power spectrum. The parameters S_{11} and S_{22} give the reflection measurements of their respective port.

Important to note, VNAs measure the voltage ratio amplitude in units of dB, given by $V = 20log(A)$. The step solving for A is suppressed.

The quality factor describes the energy loss of a resonator and is given by

$$Q = \omega_0 \frac{\text{Energy stored}}{\text{Energy lost per cycle}}, \tag{B.1}$$

where ω_0 is the resonant frequency. The higher the quality factor of a resonator, the less energy it dissipates. A resonator has two quality factors, the loaded quality factor, Q_L, and the unloaded quality factor, Q_U. The unloaded Q factor describes the inherent energy losses of a resonator based on its material composition and geometry. The loaded Q factor describes the energy losses of a resonator with a load, which will always be equal to or less than the unloaded Q factor. The loaded quality factor can be calculated using

$$Q_L = \frac{\omega_0}{\Delta\omega}, \tag{B.2}$$

where $\Delta\omega$ is the FWHM of the signal. The loaded quality factor describes the total energy loss of the resonator. For a two-port cavity, this corresponds to the inherent losses of the cavity and losses through the two ports, so it is given by

$$\frac{1}{Q_L} = \frac{1}{Q_1} + \frac{1}{Q_2} + \frac{1}{Q_U}, \tag{B.3}$$

A. Hipp, *Searching High and Low for the Axion*,
Springer Theses, https://doi.org/10.1007/978-3-032-02276-9

where Q_1 and Q_2 are the quality factors of port 1 and port 2, respectively. This can be simplified to

$$\frac{1}{Q_L} = \frac{\beta_1 + \beta_2 + 1}{Q_U}, \tag{B.4}$$

with β_1 and β_2 being the couplings of port 1 and port 2, respectively. The unloaded quality factor can be computed by using a VNA to measure Q_L and the reflection dips of the TM_{010} [1].

The VNA calculates Q_L using Eq. B.2. The TM_{010} mode, the lowest frequency mode, is typically used. The VNA gives power in units of dB, so the FWHM corresponds to points on either side of the resonance 3 dB down from the resonant peak. If a point is not exactly at 3 dB down, as is typically the case, the frequency is interpolated. A correct frequency resolution must be selected so the interpolation provides an accurate estimate.

The beta parameters are calculated using the S_{11} and S_{22} spectra. On resonance, the reflection dip for a mode yields

$$S_{110} = \frac{1 - \beta_1 + \beta_2}{1 + \beta_1 + \beta_2}, \tag{B.5}$$

$$S_{220} = \frac{1 + \beta_1 - \beta_2}{1 + \beta_1 + \beta_2}, \tag{B.6}$$

where the third subscript, zero, specifies the measurement is made on resonance. Solving for the beta parameters yields

$$\beta_1 = \frac{1 - S_{110}}{S_{110} + S_{220}}, \tag{B.7}$$

$$\beta_2 = \frac{1 - S_{220}}{S_{110} + S_{220}}. \tag{B.8}$$

Substituting the betas into Eq. B.4 and solving for the unloaded quality factor yields

$$Q_U = \frac{2Q_L}{S_{110} + S_{220}}. \tag{B.9}$$

At critical coupling, $S_{110} = S_{220} = 0.5$, so the unloaded Q is exactly double the loaded Q. For very weak coupling, $S_{110} = S_{220} \approx 1$, yielding an unloaded Q approximately equal to the loaded Q.

References

1. D.B. Tanner, Two-port cavities: coupling and cavity Q (2021), unpublished
2. D.H. Han, Y.S. Kim, M. Kwon, Rev. Sci. Instrum. **67**, 2179 (1996)

Appendix C
LC Tuning

C.1 Parallel Plate Capacitor

The LC circuit was tuned by inserting a dielectric slab into the parallel plate capacitor, as shown in Fig. C.1.

The general formula for capacitance is given by

$$C = \frac{\varepsilon A}{d}, \tag{C.1}$$

where A is the area of the parallel plate capacitor, ε is the permittivity, and d is the separation between the plates. Since the dielectric in our capacitor does not fill the entire width and will at times be only partially inserted, the cases of dielectrics in series and parallel must be considered.

For dielectrics in parallel, the total capacitance is given by $C = C_1 + C_2$, where C_1 and C_2 are the found using Eq. C.1 using the respective area and permittivity of each dielectric slab. The capacitance for dielectrics in series is $\frac{1}{C} = \frac{1}{C_1} + \frac{1}{C_2}$. Let the area of the parallel plate capacitor be given by $A = H \times L$, where H is its height and L is its length, the insertion depth of the dielectric slab be h, and the permittivity of the dielectric be $\varepsilon = \varepsilon_r \varepsilon_0$. The total capacitance for the parallel plate capacitor set-up shown in Fig. C.1 is

$$C = \frac{\varepsilon_0 (H - h) L}{d} + \frac{\varepsilon h L}{\varepsilon_r (d - w) + w}. \tag{C.2}$$

ADMX SLIC uses a parallel plate capacitor with dimensions 2" × 2" and a separation of 0.1825". A slab of alumina is used as the dielectric, which has a relative permittivity of $\varepsilon_r = 9$ and a thickness of 0.06".

A. Hipp, *Searching High and Low for the Axion*,
Springer Theses, https://doi.org/10.1007/978-3-032-02276-9

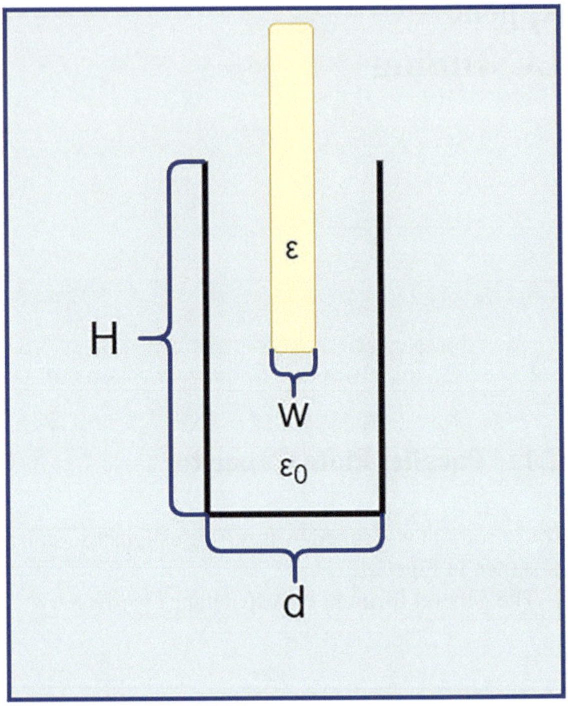

Fig. C.1 Parallel plate capacitor: height H, separation d; dielectric slab: permittivity ε, width w

C.2 Inductance

The LC circuit uses a single rectangular loop of NbTi wire with dimensions 3" × 12.5" and a diameter of 0.25 mm. The pickup loop is also 0.25 mm diameter NbTi wire, and is a circular loop with 5 turns and a 2 cm diameter. The self-inductance in units of μH of a rectangular loop of wire with side lengths a and b, diagonal $c = \sqrt{a^2 + b^2}$, and diameter d given in inches is [1]

$$\begin{aligned} L =& 0.02339[(a+b)\log_{10}\frac{4ab}{d} - a\log_{10}(a+c) - b\log_{10}(b+c)] \\ &+ 0.01016[2c + d - 2(a+b)]. \end{aligned} \quad \text{(C.3)}$$

The self-inductance is found to be $L = 0.968$ µH. The pick-up loop used for inductive coupling utilizes the same NbTi wire and has $n = 5$ turns with a $D = 2$ cm diameter. The self-inductance in units of μH and all dimensions in inches can be found using [1]

$$L = 0.1595 D n^2 [2.303 \log_{10}\frac{8D}{d} - 2]. \quad \text{(C.4)}$$

The self-inductance of the pick-up loop is found to be $L_2 = 1.4\ \mu$H.

C.3 LC Circuit Frequency

The resonant frequency of an LC circuit is given by

$$\omega = \frac{1}{\sqrt{LC}}. \tag{C.5}$$

The inductance of the LC circuit was affected by environment parasitic inductances. A measurement of the frequency and a calculation of the capacitance allowed us to use Eq. C.5 to deduce a total inductance of $L = 1.8\ \mu\text{H}$. The frequency of the circuit as a function of depth of the dielectric slab in the parallel plate capacitor is shown in Fig. C.2. Future runs may use an alumina slab with a 0.1" thickness, instead of 0.06". Figure C.3 shows the tuning range using 0.1" thick slab of alumina.

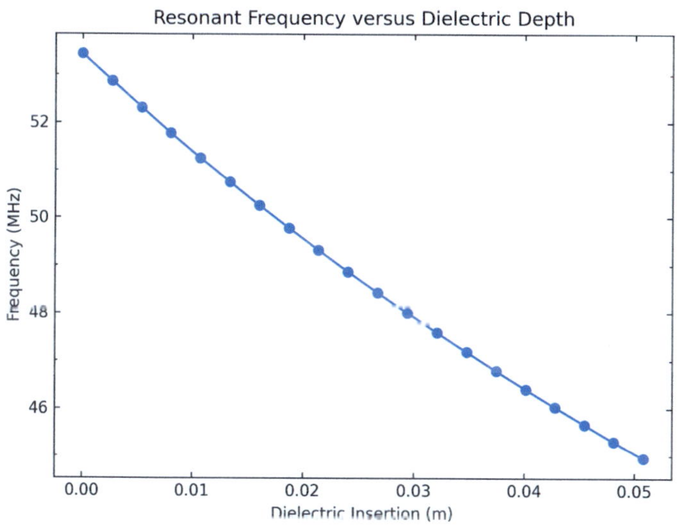

Fig. C.2 Resonant frequency of the LC circuit with respect to the alumina slab depth

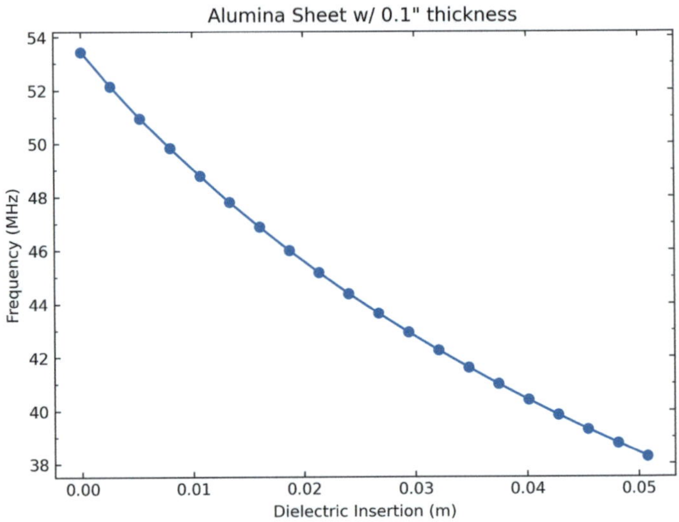

Fig. C.3 Resonant frequency of the LC circuit with respect to the depth of an alumina slab, with 0.1" thickness

Reference

1. F.E. Terman, *Radio Engineers' Handbook*, 1st edn. (McGraw-Hill Book Company Inc, London, 1943)

Curriculum Vitae

Alexander Hipp
POSTDOCTORAL SCHOLAR
2233 Tech Dr., Mudd Rm 1185, Evanston, IL 60208
e-mail: alexander.hipp@northwestern.edu | alexanderthipp

Education

University of Florida *Gainesville, FL*
PH.D. PHYSICS *2024*

- Advisor: David B. Tanner
  Thesis: Searching high and low for the axion

University of Florida *Gainesville, FL*
M.S. PHYSICS *2022*
Edgewood College *Madison, WI*
B.S. PHYSICS & MATHEMATICS *2019*

A. Hipp, *Searching High and Low for the Axion*,
Springer Theses, https://doi.org/10.1007/978-3-032-02276-9

Professional Experience

Postdoctoral Scholar, *Northwestern University* *Evanston, IL*
ADVISOR: ANDREW GERACI *Aug 2024-Present*

- Lead for construction and commissioning of ARIADNE detector
- Designs and machines detector components

Graduate Research Assistant, *University of Florida* *Gainesville, FL*
ADVISOR: DAVID B. TANNER *May 2020-May 2024*

- Performed high resolution data analysis for the ADMX Run 1c data set
- Characterized the performance of microwave cavities for future axion searches
- Developed data acquisition system for ADMX SLIC

Graduate Teaching Assistant, *University of Florida* *Gainesville, FL*
PHY2048L & PHY2053L *Aug 2019-May 2020*

- Taught lab sections for algebra and calculus based introductory physics

IREU Undergraduate Researcher, *University of Glasgow* *Glasgow, UK*
ADVISOR: IK SIONG HENG *Jun-Aug 2018*

- Conducted research associated with LIGO and VIRGO
- Simulated seismic wave backgrounds near gravitational wave detectors
- Computed strain estimates due to seismic Newtonian noise

Awards and Fellowships

2023 **E. Raymond Andrew Memorial Award**, University of Florida, Department of Physics
2020 **IHEPA Graduate Fellowship**, University of Florida, Department of Physics
2019 **Outstanding Graduating Senior Physics Student**, Edgewood College

Talks

APS April Meeting 2024 *Sacramento, CA*
ADMX Run 1c: A High-Resolution Search *Apr 2024*
The Early Universe: A Window to New Physics Workshop (Invited) *Gainesville, FL*
ADMX Run 1c: Multi Resolution Analysis *Oct 2023*
UF High Energy Seminar (Invited) *Gainesville, FL*
ADMX: An Overview *Sep 2023*
APS April Meeting 2023 *Virtual*
ADMX: Run 1c HiRes Analysis *Apr 2023*
ADMX Collaboration Meeting *Seattle, WA*
HiRes Analysis: Overview and Current Work *Jan 2023*

Snowmass 2021 Poster Session *Seattle, WA*
ADMX Run 1c HiRes Analysis *Jul 2022*
ADMX Collaboration Meeting *St. Augustine, FL*
HiRes Analysis: Persistence. *May 2022*
APS April Meeting 2022 *Virtual*
ADMX Extended Frequency Range: 2-4 GHz Search *Apr 2022*
ADMX Collaboration Meeting *Seattle, WA*
HiRes Analysis of the ADMX Run 1c Data *Aug 2021*
16th Patras Workshop on Axions, WIMPs and WISPs *Virtual*
HiRes Analysis of the ADMX Run 1c Data *Jun 2021*
APS April Meeting 2021 *Virtual*
ADMX High Resolution Analysis *Apr 2021*

Publications

PUBLISHED

[1] M. Guzzetti et al., "Improved receiver noise calibration for ADMX axion search: 4.54 to 5.41 μeV", Phys. Rev. D **111**, 092012 (2025).
[2] C. Goodman et al., "ADMX Axion Dark Matter Bounds around 3.3 μeV with Dine-Fischler-Srednicki-Zhitnitsky Discovery Ability", Phys. Rev. Lett. **134**, 111002 (2025).
[3] C. Bartram et al., "Non-Virialized Axion Search Sensitive to Doppler Effects in the Milky Way Halo", Phys. Rev. D **109**, 083014 (2024).
[4] S. Chakrabarty et al., "Low Frequency (100-600 MHz) Searches with Axion Cavity Haloscopes", Phys. Rev. D **109**, 042004 (2024).
[5] C. Boutan, et al., "Axion Dark Matter eXperiments: Run 1A Analysis Details", Phys. Rev. D **109**, 012009 (2024).
[6] T. Nitta et al., "Search for a Dark-Matter-Induced Cosmic Axion Background with ADMX", Phys. Rev. Lett. **131**, 101002 (2023).
[7] C. Bartram et al., "Dark Matter Axion Search Using a Josephson Traveling Wave Parametric Amplifier", Review of Scientific Instruments **94**, 044703 (2023).
[8] C. Bartram et al., "Search for Invisible Axion Dark Matter in the 3.3–4.2 μeV Mass Range", Phys. Rev. Lett. **127**, 261803 (2021).

IN REVIEW

[1] G. Carosi et al., "Search for Axion Dark Matter from 1.1 to 1.3 GHz with ADMX", arXiv:2504.07279 (2025).
[2] A. T. Hipp et al., "Search for non-virialized axions with 3.3–4.2 μeV mass at selected resolving powers", arXiv:2410.09203 (2024).

Skills

Technical	Cryogenic methods, machining, microwave cavity S-Parameter measurements
Data analysis	Fourier methods, SNR estimates, probability theory
Programming	Python, SQL, bash, LaTeX, HTML

Mentoring

Jack Ryan, Undergraduate Researcher, Northwestern University *2024–Current*
Mitchell Solano, Graduate Research Assistant, University of Florida *2024–Current*
Thomas Caligiure, Undergraduate Researcher, University of Florida (Now a graduate student at University of Florida) *2021–2024*

MIX
Papier aus verantwortungsvollen Quellen
Paper from responsible sources
FSC® C105338

If you have any concerns about our products,
you can contact us on
ProductSafety@springernature.com

In case Publisher is established outside the EU,
the EU authorized representative is:
Springer Nature Customer Service Center GmbH
Europaplatz 3, 69115 Heidelberg, Germany

Printed by Libri Plureos GmbH
in Hamburg, Germany